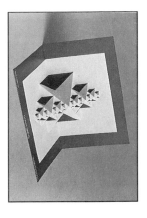
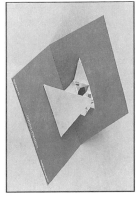
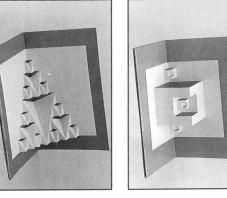
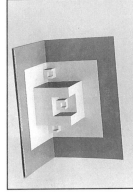
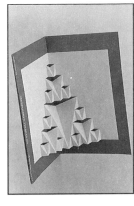

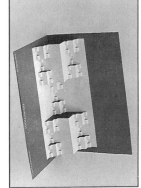

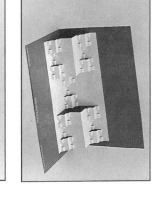
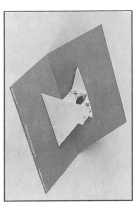

FRACTAL CUTS

*Exploring the magic of fractals
with pop-up designs*

◆

Diego Uribe

Tarquin Publications

W9-BAB-514

Diego Uribe is an industrial designer of international repute with a lifelong interest in mathematical ideas. Based in Buenos Aires, he has been writing articles on recreational mathematics for newspapers and magazines in Argentina and Spain for the last fifteen years.

An early exponent of computers when they were anything but user-friendly, he is an experienced programmer who appreciates the power of computer-aided design. Fractals have been a long term interest.

This book on fractal cuts is the culmination of his enthusiasm. It offers wonderful designs based on simple but not trivial mathematics and demonstrates the imaginative use of computers to calculate and illustrate the results.

However, this marvellous visualisation of fractal ideas requires neither drawing skills nor the use of a computer. Patience and a craft knife are all that is needed to create these pop-up versions of endlessly fascinating fractals.

An up-to-date catalogue can be obtained from the publisher at the address below.

© 1995:   Diego Uribe   Tarquin Publications
© 1993:   Previous Edition   Stradbroke
I.S.B.N.: 0 906212 88 X   Diss
Design: Magdalen Bear   Norfolk IP21 5JP
Cover: Paul Chilvers   England
Printing: Foister & Jagg Ltd.

7. Fanned Triangles

5. Tetrahedron Pairs

1. Central Quartiles

2. Trisections

8. Opposed Pyramids

9. Disappearing Vertices

"*Endlessly developing pop-up designs which belong to the class of curious geometrical objects called fractals.*"

◆

3. Alternate Steps

6. Sierpinski's Triangle

4. Contrary Cubes

10. Hanoi Arête

Curious geometrical objects

This illustration does not seem to be of a curious geometrical object at all, rather more a kind of weed, perhaps blown by the warm winds of some unknown tropical plain. Examine its detail: an intricate web constructed from hundreds of minute stems. It seems that the artist certainly had a hard time creating such a complex drawing!

Intricate? Complex? Well, not quite. Have a further look. Can you see that the whole weed is made of copies of a single motif, a rhombus with three lines sprouting from some corners?

No? Well, let's try to draw it.

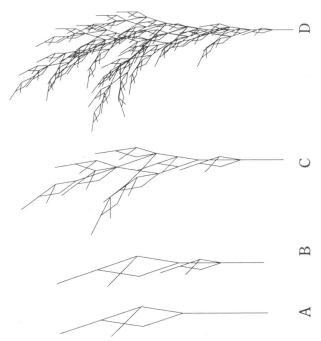

A   B   C

D

Let us start with a very simple drawing (A), made of eight segments, all of equal length. The four sides of the rhombus are equal, as are the two extra lines sprouting from it. The vertical line is made of two more. This is our basic motif, and we shall draw it time and again.

The next step is to replace each of the eight segments of this motif by a smaller copy of itself. In fact, one which is exactly half the size. Drawing B shows what happens when we replace the lower vertical segment with a half size copy of the whole motif. Drawing C shows what happens when we then repeat the same process on the rest of the segments.

Now, the smaller motifs are also made of eight equal segments, so the next step is to replace each of them by a half size copy of the smaller motif (D). These new motifs are of course a quarter the size of the original.

One step more and we are back at the tropical weed. The illustration wasn't so intricate after all, just mechanical repetition of a simple motif over and over again.

Endless repetition and infinite detail

It is clear that we can continue the procedure for ever, drawing smaller and smaller details. The only limits are the thickness of the pen trace, our drafting skill or simply the lack of infinite time to spend on it. Of course, real weeds do not have infinite detail; there will always be a stem or a leaf that is the smallest. However, nothing prevents us from imagining, not a real object, but an ideal one; a mathematical object, a fractal.

The word fractal was coined by the French mathematician Benoit B. Mandelbrot. This weed is an example of a regular fractal, one in which all the segments are of equal length and in which all are replaced by a smaller copy of the complete motif. Each repetition of the procedure produces a new 'generation' and all the motifs of a particular generation are the same size.

Let us now examine a second procedure and a second fractal.

Divide a line into three equal parts.

Replace the central section with two sides of an equilateral triangle.

Replace each segment with a complete motif, one third of the size.

Plainly this procedure could be continued for ever, giving more and more detail. This method of replacing a segment with a smaller motif made of segments and then replacing each of those segments with yet smaller motifs is referred to in several different ways in different circumstances. Sometimes it is called the 'segment and motif' method, sometimes the 'draw and replace' method and sometimes the whole procedure is loosely referred to as a 'generating rule'. There does not seem to be any confusion once the concept is understood and all uses will be found in this book.

Drawing a snowflake

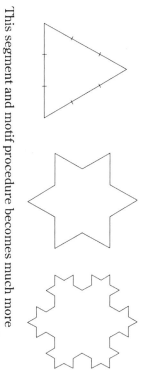

This segment and motif procedure becomes much more interesting when it operates on an equilateral triangle. When all three middle segments are converted to outward pointing equilateral triangles, each one third the size of the original, the result is a six-pointed star with twelve sides.

Repeating the procedure gives an eighteen pointed star with forty eight sides.

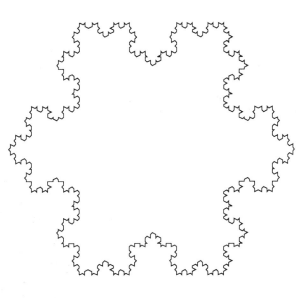

After a few more stages the figure does begin to resemble a snowflake. However, we have not reached the end of the story. The process can continue for ever, stage after stage, always obeying the same rule; each time replacing the middle part of the sides with outward pointing equilateral triangles.

Self-similarity

The 'mathematical snowflake' which we have drawn overleaf, has remarkable properties. Consider the border that goes from one corner of the original triangle to another. Just like the fractal weed, it is made of reduced copies of the whole border. Or to put it another way, any portion, provided we enlarge it by the proper factor, is identical to the whole border. This property is called self-similarity.

The snowflake was discovered at the beginning of this century by the German mathematician Helge von Koch. One of its earliest investigators was the Italian mathematician, Ernesto Cesaro, who wrote:

"It is this self-similarity in all its parts, however small, that makes the curve seem so wondrous. If it appeared in reality, it would not be possible to destroy it without removing it altogether. For otherwise it would ceaselessly rise up again from the depths of its triangles like the life of the universe itself."

Cesaro has also proposed an alternative way of constructing it.

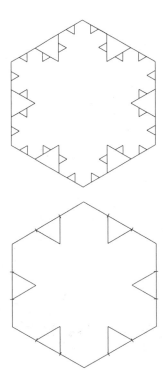

This time let us start with a hexagon and divide each of its sides into three equal segments. We then cut equilateral triangular pieces from the centre section of each side. The diagram above shows the first two generations and it can be seen that after an infinite number of generations this procedure will produce the same snowflake design.

This illustration shows both methods at the same time. The hexagon is drawn in black over a tinted background and the inner triangle is drawn in white. As we eliminate black triangular pieces from the hexagon and add white ones to the triangle, the remaining black area begins to shrink, until after an infinity of generations it vanishes completely and the two outlines coincide.

These methods of construction also throw light on another curious property of the snowflake. Whichever method we use, the perimeter of the snowflake increases steadily with every generation. Every time we add a new triangle in one case or subtract one in the other, the border becomes longer by one unit. In the limit it therefore becomes infinite. On the other hand, the area of the snowflake is not infinite. Its value lies somewhere between the areas of the inner triangle and of the outer hexagon.

So we have a magical shape that can be reconstructed from the tiniest portion of its border and which has a finite area bounded by a border of infinite length! No wonder when the first examples of such figures were discovered in the last half of the last century, many mathematicians called them 'monsters' or even 'pathological cases'!

Infinite perimeters

The snowflake fractal is a figure with an infinite perimeter and a finite surface area. We are now able to introduce one with an infinite perimeter, but with no surface area at all!

Let us start with a triangle, divide it into four equal half-sized triangles and remove the central one.

Since we are concerned with the perimeter, this diagram shows a slight modification so that the border of the figure can be clearly drawn with a continuous line.

It is then clear that this shape has a perimeter which is four times the side of the original triangle.

The next generation is obtained by dividing each of the three remaining triangles into four and eliminating all their centres. It is clear that the perimeter is increased by nine quarters of the original side and the surface area reduced by three sixteenths.

We then continue with the procedure generation by generation, adding to the perimeter and reducing the area. After an infinite number of generations we end with a closed, infinite line, encircling ..... nothing!

This fractal is known as Sierpinski's Gasket after the mathematician W. Sierpinski who first described it in 1915. This illustration shows a Gasket with five generations. There are three times as many holes in each successive generation and they are each half the size. By counting the number of different sized holes present, it is a simple matter to identify how many generations it has.

Sierpinski's carpet

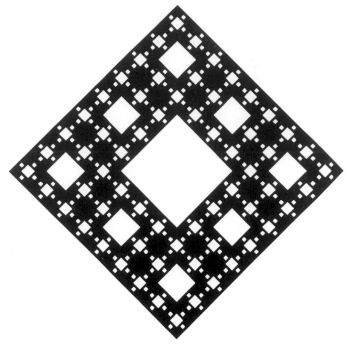

This example shows another curious property of some fractals. Let us start with a segment and replace it with this eight-segment motif; a kind of calculator display zero with two additional segments sprouting from its sides.

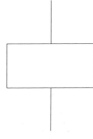

Notice that this motif has two-fold symmetry. That is, if you spin it around its central point, it will coincide exactly with itself twice every complete turn. Once, after a rotation of 180° and again after a rotation of 360°.

Let us now begin the construction of the fractal, generation by generation, each time replacing every one of the eight segments by a one-third sized copy of the complete motif. It is clear that the next two generations also have two-fold symmetry.

However, notice how the 'holes' in the pattern seem to be getting squarer as the fractal develops. Is it possible that in the limit this fractal could have four-fold symmetry?

This illustration does seem to suggest that in the limit the holes do become squares and that a fractal with four-fold symmetry can be generated from a motif which has two-fold symmetry. However, it was obtained by a quite different procedure.

It started with a square divided into nine smaller squares and the central one was removed. Each of the eight remaining squares was replaced by a complete motif one third of the size.

After a number of generations the image above was the result.

A square has four-fold symmetry and this is unchanged by removing its centre. Hence the complete fractal must have four-fold symmetry.

If we now look at both types of illustration from sufficient distance so that the details become blurred, both methods do seem to produce the same fractal, a fractal with four-fold symmetry. It is known as the Sierpinski carpet.

Peano's curve

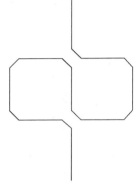

Let us now meet another curious property of fractals. This time we start with a nine-segment motif, similar to the one opposite, but with an extra line across the centre. Since the property concerns a continuous line, the corners have been chamfered a little to make it clearer to see. Once again, generation by generation, we replace every segment of the motif by a one-third sized copy of itself.

After just one generation the pattern looks like this. Note that the curve remains a continuous one. From start to finish there would be no need to lift the pen from the paper.

After only a few steps the pattern becomes more and more detailed, but the curve never ventures outside a certain square. However, it will eventually pass arbitrarily close to every possible point within that limiting outline. After an infinity of generations, the curve will therefore completely fill the square.

Elementary geometry teaches us that a line has only one dimension, i.e. its length, whereas a square has two, i.e. length and width, and a solid has three, i.e. length, depth and width. This type of curve is truly a paradox. We seem to have constructed a line, a one-dimensional object, that is completely indistinguishable from a square, a two-dimensional object!

This curve was discovered by the Italian mathematician Guiseppe Peano, and although it is not a true fractal in a strict mathematical sense (the reason is quite technical), it shares many of the properties of fractals, including the way it is constructed. Such space-filling curves are known as Peano curves. What then is their true dimension? Since a Peano curve is really a way of looking at a piece of a plane, it should be of dimension 2. However it is also a one-dimensional line.

To overcome this conflict mathematicians have introduced the idea of a 'fractal dimension'. An intuitive definition is that the fractal dimension describes the amount of the plane that the curve eventually fills. A Peano curve fills the whole plane and therefore has a fractal dimension of 2. The Koch curve on page 5 does not fill the plane although it covers more of it than a straight line. Its fractal dimension therefore lies between 1 and 2. It is in fact 1.26182. The curve on page 40 has a fractal dimension of just over 1.8. This topic is dealt with in more advanced texts and is one which repays further study.

Paper dragons

One of the simplest and most surprising methods of generating a fractal is suggested by folding a strip of paper.

Fold it in half three times, always in the same direction.

Then open it so that every fold is a right angle and it will assume the eight-segment shape shown here.

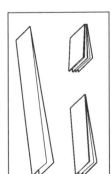

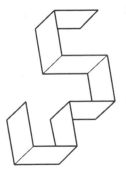

The edge of the paper is a self-similar curve with infinite detail, granted that the method of folding can be applied to each segment and therefore that the folding can go on for ever. This, and other closely related fractals, are known as paper dragons and were discovered by the American physicist, John E. Heighway.

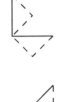

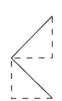

The paper dragon can also be obtained by the draw and replace method. Start with a segment and replace it with a right angled isosceles triangle. Now, imagine a journey along the two equal sides, moving in a clockwise direction. To construct the next generation, replace the first segment by a smaller right angled isosceles triangle on your left and the second segment by a smaller right angle placed at your right.

As this procedure continues it will be observed that some points would coincide and the detail would become blurred. To avoid this, we have replaced the sharp right angles with small circular arcs.

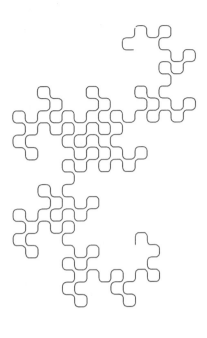

An interesting property of the paper dragon is that four copies placed like the vanes of a windmill completely cover the plane. Like the Peano curve, its fractal dimension is 2.

Draw and replace

All four fractal curves on this page were constructed by using the draw and replace method. Can you deduce how they were created?

Cesaro was the author of the thorned fractal immediately below and the American mathematician, William A. McWorter Jr. created the pentagon fractal beneath it.

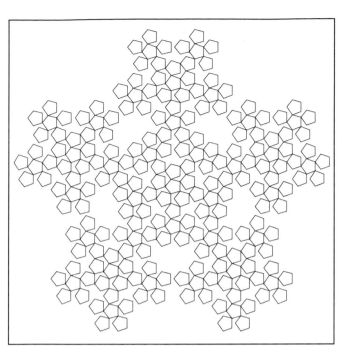

This Peano curve is by W. Gosper. Interestingly, a portion of its interior matches that on an old Inca textile. Unfortunately the piece of cloth is not big enough to see whether the design follows a true self-similar curve or just a simple repetitive pattern.

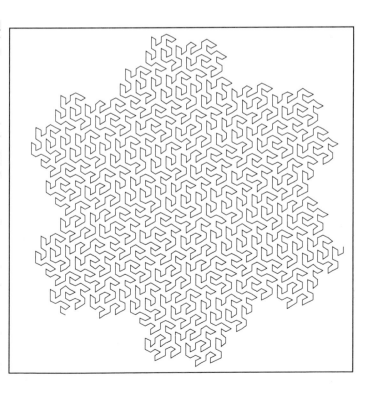

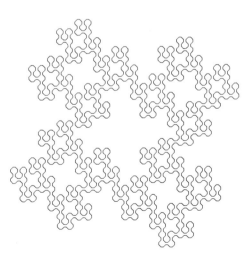

There are still other ways of producing fractals on computers with graphic displays, without resorting to probability and chance. They bear names as strange as 'iteration function systems' or 'domains of attraction'. The fractals produced in this way have infinite detail and self-similarity but in most cases distortion is introduced. The details are distorted copies of the whole object, although distorted according to a precise mathematical procedure. For instance, a vertical segment can be sheared, curved or rotated, generation by generation, but always by the same amount.

Both illustrations on this page are computer generated fractals. The black fern leaf on the left is due to Michael Barnsley, an American computer scientist, who also developed the program to generate it. Notice that every part reproduces the complete leaf.

Below is a 'computer dragon' with a border very similar to the paper dragon, although produced using a similar program to the one for the black fern.

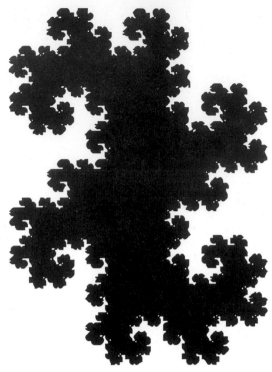

There are many books and computer programs of fractals now available, containing or generating images of bewildering beauty and apparent complexity.

Computer generated fractals

Since fractals are obtained generation by generation by the repeated application of a simple rule, it is a subject which lends itself remarkably well to the use of computers. In fact, it is the availability of fast, powerful computers with good graphical displays which has so stimulated interest in the subject in recent times.

So far each of the fractals we have met in this book has been regular, in the sense that any portion is identical to the whole, once enlarged by a suitable multiplying factor. However, fractals can also be self-similar in a less precise, statistical way. Suppose, for instance, that we want to construct a paper dragon but decide to introduce a slight modification to the generating procedure. Instead of rigidly placing the right isosceles triangles alternately left and right of the line of travel, we toss a coin to determine which side to place it. If we throw heads, then we place it on the right side; if tails, on the left side.

The resulting fractal will have the same number of units as the original paper dragon, but they will no longer be regularly arranged. It will still be self-similar in a statistical sense because any portion will, in the limit, have the same number of triangles on the right as on the left as the complete curve. As the probability of getting heads when you toss a coin is one half, half of the angles will be right-handed and half will be left-handed. Moreover, as the curve is infinitely complex, the smaller portion will resemble, except for the detail, the whole fractal.

12

Creating complex images

Once Mandelbrot and others had produced some stunning images using fractals, many people were attracted to the idea of fractal art, especially working in colour. The image above, called 'Whirlpools', was generated by a procedure similar to the one used to produce the famous Mandelbrot Set, a fractal object so popular that it has been used on many postcards and T-shirts. Notice that each of the spirals is made from lesser spirals. Those are of course in turn made of still smaller spirals and so on generation by generation, down to the infinitely small. If we now browse over this illustration and the two previous ones, it is evident that these examples of computer generated fractals are far better representations of natural phenomena than any of the regular fractals seen earlier. The whirlpool design does seem to suggest real turbulent water. The fern looks almost real and is far more naturalistic than the tropical weed introduced at the beginning of the book. In fact, fractals can be designed to model almost any shape we can imagine.

George Lucas, the director of films such as Star Wars, has used fractals to simulate the skies, mountains and vegetation of alien landscapes. Not only do they look sufficiently strange to add credibility to the story but they can be generated relatively cheaply. There is no need to construct huge and expensive sets and of course changes can be made very easily, simply by changing the generating rule.

Another important modern use of fractals is for the compression of large computer images. Let us consider, for example, the photographs which we took on our last holiday trip. A good part of each image is occupied by the landscape: beaches and seas, hills and mountains, plains and woods. Probably some part of each scene will be sky. Only in the foreground is the focal subject of our snapshots: our friends or families. Now, let us suppose we want to display one of these photographs on a computer display or to use a computer to manipulate or print it. Every single point of the photograph must be translated into computer code indicating its position, hue, brightness and saturation. Even a small, low resolution computer display needs more than a quarter of a million of such points. The amount of information which needs to be stored for any image is indeed huge.

Michael Barnsley, the author of the black fern fractal, was one of the first to devise a method of using fractals to compress such images. The photograph is scanned to detect elements which could be represented by fractal structures in the background. An automatic procedure to model those fractals is set up and the enormous quantity of information needed to display the background can be replaced by relatively few short instructions, thus dramatically reducing the size of the image file.

The image displayed on the computer will then be a mixture of the actual details of our friends in the foreground and fractal reconstructions of parts of the background. Of course there will be some differences. Although the trees will look like the oaks or the maples which were actually there, and although the clouds in the image will look like the clouds in the real sky, the details will not match exactly the photograph we took. Say, this leaf will be somewhat smaller or greener and that rock will have more cracks than the original. However, the scene as a whole will closely match the original one. The benefits and convenience of handling images by computer are very great and it is certain that fractals will play an ever greater role in the future.

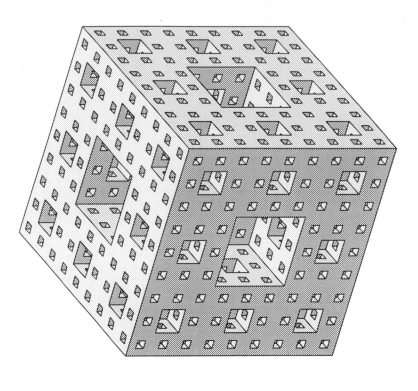

It is also easy to imagine a solid version of the Sierpinski carpet. Start with a cube and, instead of removing squares, remove cubes. The result is known as the Sierpinski sponge and it is a solid bounded by six Sierpinski carpets.

All these three-dimensional objects share the strange properties of their plane relatives. For example, as the Sierpinski sponge endlessly develops, the perimeter surrounding its holes grows to infinity while its total volume vanishes to zero.

Fractals in space

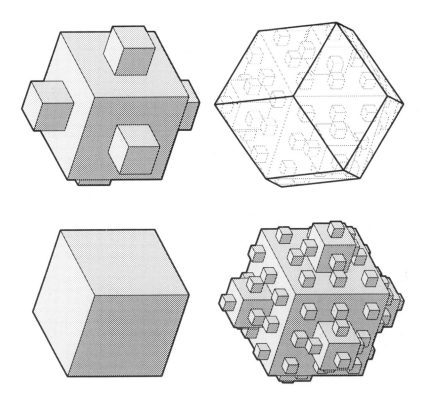

These illustrations suggest that fractal objects are not limited to two dimensions and the plane, but exist in space as well.

For example, we can easily adapt a square version of the snowflake curve procedure to get a fractal surface. Instead of starting with a triangle, let us start with a cube. Instead of replacing the middle segment of each side of the triangle with a smaller triangle, let us divide each face of the cube into nine squares and build a one-third sized cube upon the central one. The illustration above shows two generations and also indicates what happens in the limit. After an infinity of generations, the exterior will take on the shape of a rhombic dodecahedron, a polyhedron bounded by twelve rhombic faces.

Fractal cuts

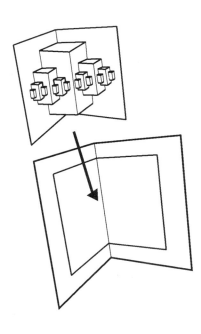

Each of the pop-up cards also produces a three-dimensional manifestation of fractals, but in a very different sense to the space-filling fractals described opposite. Since fractal cuts have to fold flat and to pop up when the card is opened, it is an interesting task to search for fractal designs which can be adapted or modified to have this property.

As is so often the case with creative work, the restraints and restrictions which are imposed upon designs by such choices of techniques and materials, add rather than detract from the beauty and interest of the finished result.

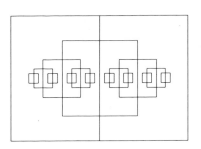

Apart from their fractal qualities, a satisfying feature of fractal cuts is that they are made from a single piece of paper, with nothing added and nothing taken away.

In regular fractals, all segments of the motif are replaced by a reduced copy of the complete motif. It is a feature of those designs which can be used for pop-up cards that only certain segments can be replaced. These segments are known as the 'active lines' and they are marked with black diamonds.

active line

Fractals in general can be described by means of a 'generating rule' which acts as a kind of instruction set for producing one generation from another. Such instructions may be expressed in words, but they are essentially a kind of geometrical procedure which has to be followed. Often the instructions include a figure from which to start. Fractal cut 9 is a good example of such a procedure. It is easy to comprehend but hard to describe in words.

Since the motifs used for fractal cuts must generate figures which pop up and fold flat, they have to have a central line of symmetry. The best method of description for pop-up motifs is in terms of a line called 'the initial line'. This initial line provides the first fold and it is normally divided in some fixed ratio. It is a convenient way of providing the framework on which the motif is drawn.

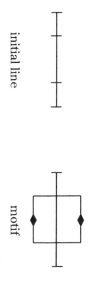

initial line

motif

Scale Factor 1:2   Multiplication Factor 2:1

The scale factor indicates that each motif in a generation is half the size of the previous one. The multiplication factor indicates how the number of motifs increases generation by generation. It is of course the number of active lines which determines the multiplication factor.

The search for beauty

Since fractals are produced by the mechanical repetition of a generating rule, it is easy to lose sight of the fact that choosing the right design and the right proportions for it is of the greatest importance. Human judgement is absolutely essential, as is patience and the willingness to spend time on trial and error.

We must also bear in mind that a true fractal is an object with infinite detail and since infinite time for drawing or folding it is not available, our models, just like the weed or the paper dragon drawings, must fall short of it. The choices of how many generations to include, what proportion of the cover the fractal cut should occupy and whether or not the base should be trimmed, are all matters on which to exercise judgement.

The remainder of this book is devoted to ten designs which do make successful pop-up cards. When they are completed they may be kept as a three-dimensional encyclopaedia of fractal ideas or simply used as attractive cards to send to friends.

Apart from the physical pleasure of making them and experiencing in a practical way an understanding of how fractals develop, each of the designs allows us to introduce more ideas about fractals and to gain more insights into their remarkable properties.

FRACTAL CUTS

Card 1. Central Quartiles

◆

The cuboids of successive generations of this fractal halve in the size and double in number as they move inwards towards a central line.

Although this is the simplest of all the cards to cut and fold, it demonstrates all the curious properties of fractal objects, including self-similarity and infinite detail.

Of course, no real card can show infinite detail, but even with only four generations it is clear that the design could continue for ever and that with a suitable enlargement factor, any cuboid with its successive generations is identical to the whole fractal.

Central Quartiles

From fractal to fractal cut

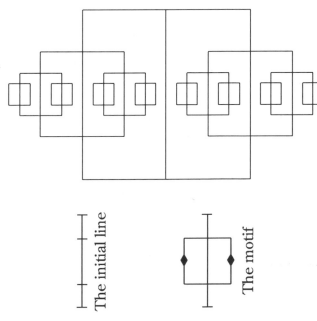

The initial line

The motif

The initial fold

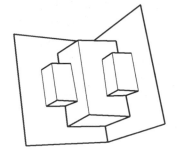

The first generation

After two generations

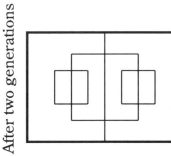

and so on...

This diagram shows the four generations of the fractal which have been used to make the pop-up card, together with the initial line and the motif which generates it.

The initial line is divided in the ratio 1:2:1 and the motif developed with a square on the central section.

The motif has two active lines and the next generation is produced by replacing each of the active lines with a complete motif, one half its size.

| Scale Factor 1:2 | Multiplication Factor 2:1 |

To convert this geometrical figure into a fractal cut, all lines at right angles to the initial line become cuts. All the other lines are scored and folded.

Work generation by generation and use the diagrams on the right to determine which folds are hill folds and which are valley folds in the finished card.

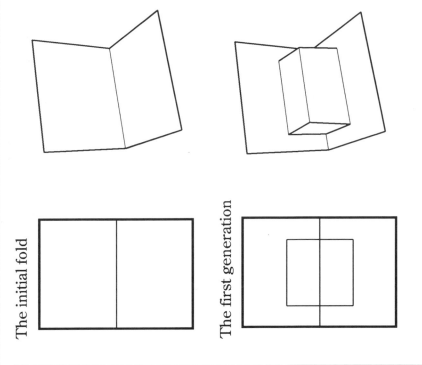

How to make
POP-UP CARD 1

This card consists of two parts, the fractal cut on this page and the cover on page 77.

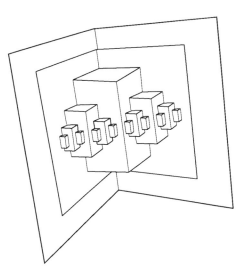

1. Remove both pages from the book and then cut out the pieces precisely.

2. Score along the fold lines and then cut along the cut lines with a scalpel.

---------------- score line

——————— cut line

3. Fold and crease the fractal cut, one generation at a time, using the diagrams opposite for guidance. Bear in mind that the white side will be visible when the card is complete. This should help you decide which are hill folds and which are valley folds. Check that at each generation the pop-up feature folds entirely flat.

4. When the fractal cut is completed, spread glue on the shaded areas at the back and glue them to the inside of the coloured cover, smoothing outwards from the central fold.

Central Quartiles

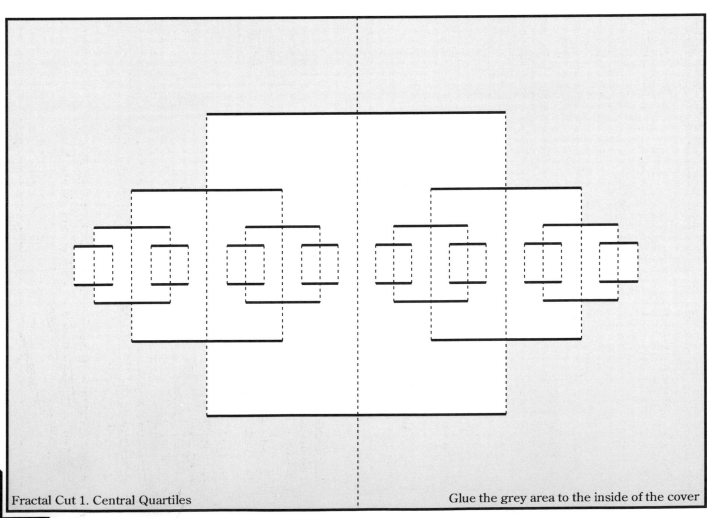

Fractal Cut 1. Central Quartiles

Glue the grey area to the inside of the cover

Further ideas

An extra delight of fractal cuts is that they contain other fractal objects within themselves.

For instance, both the pattern of cuts and the pattern of folds are fractal designs or have fractal-like qualities.

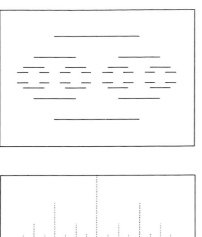

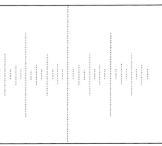

It is suggested that in making the fractal cuts, each is folded and creased one generation at a time. Not only does taking this advice emphasise the endlessly developing nature of all fractals, but one can see that the successive stages of folding also generate a sequence of fractal shapes.

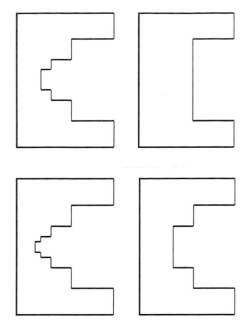

A characteristic of three-dimensional fractal objects is that a symmetrical section of them is also a fractal object.

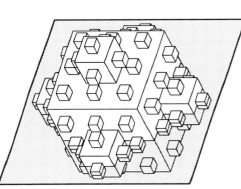

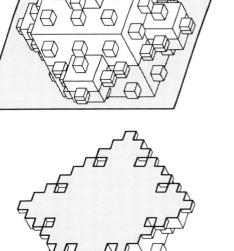

Fractal cuts have the same property, but this time we do not need to cut them. All we need to do is to look from the side.

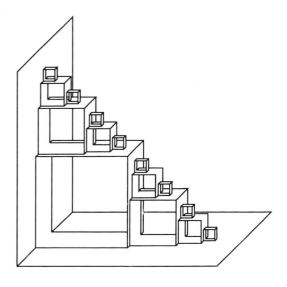

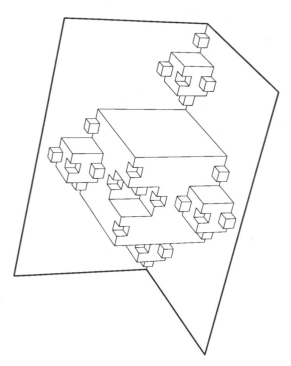

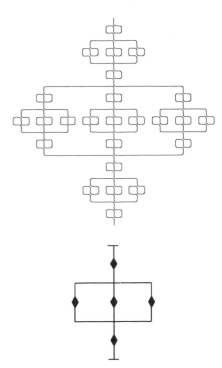

Since all motifs which can be used to create pop-up elements must fold along a central line, a motif which is square produces cuboids which have sides in the proportion of 2:1.

To produce a pop-up cube it is necessary to start with a rectangular motif which is made of two squares. This is essentially the design which creates Peano's curve, described on page 9.

However the same motif can be modified further if the central segment is also made into an active line. This is as close as we can get to Peano's curve, because in designs which pop up, vertical segments are never active lines.

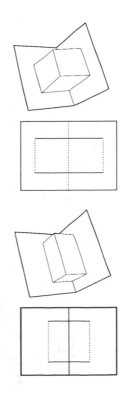

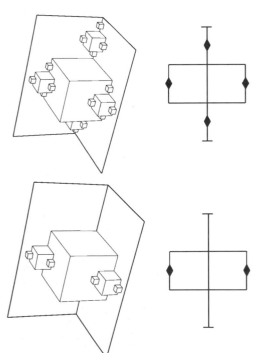

These illustrations show that the choice of which segments of a motif to make into active lines exacts a profound influence on the nature of the resulting fractal cut. Although neither is really quite interesting enough to become excited about. There seems to be too much space around the elements.

These changes mean that the generating rule also has to be modified and development then takes place along both hill and valley folds. The finished design now has cubical recesses and the effect is such as to make a very interesting pop-up card.

FRACTAL CUTS

Card 2. Trisections

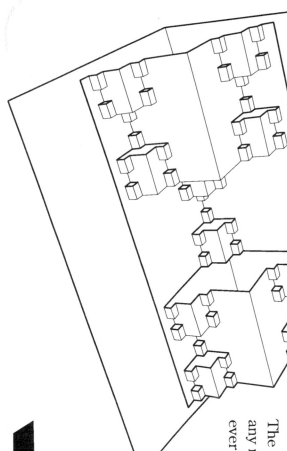

◇

With each new generation of cuboids, two thirds of the lengths of all valley folds are converted into hill folds.

It is undoubtedly true that once a hill fold has been created, it remains unchanged in future generations. Does this not mean that all valley folds will disappear and that the whole fractal will contain only hill folds? The card only shows three generations, but is there any reason why the fractal design cannot continue for ever until all the valley folds have disappeared?

Trisections

The initial fold

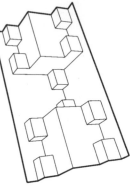

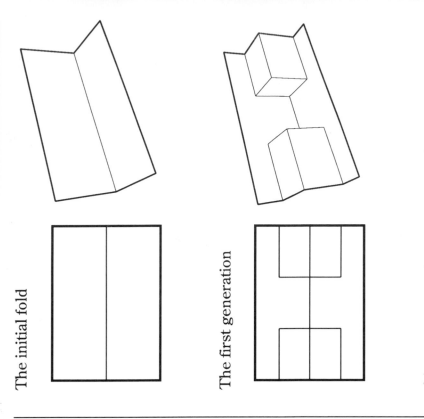

The first generation

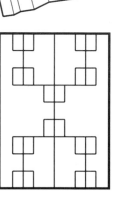

After two generations

and so on...

From fractal to fractal cut

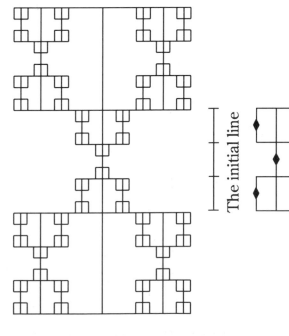

The initial line

The motif

This diagram shows the three generations of the fractal which have been used to make the pop-up card, together with the initial line and the motif which generates it.

The initial line is divided into thirds and the motif developed from two equal rectangles on the outer two.

The motif has five active lines and the next generation is produced by replacing each of the active lines with a complete motif, one third its size.

| Scale Factor 1:3 | Multiplication Factor 5:1 |

To convert this geometrical figure into a fractal cut, all lines at right angles to the initial line become cuts. All other lines are scored and folded.

Work generation by generation and use the diagrams on the right to determine which folds are hill folds and which are valley folds in the finished card.

How to make
POP-UP CARD 2

This card consists of two parts, the fractal cut on this page and the cover on page 79.

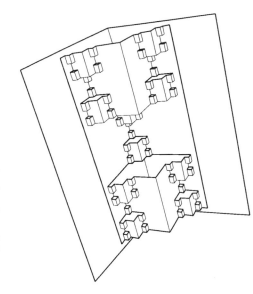

1. Remove both pages from the book and then cut out the pieces precisely.

2. Score along the fold lines and then cut along the cut lines with a scalpel.

------------- score line

———————— cut line

3. Fold and crease the fractal cut, one generation at a time, using the diagrams opposite for guidance. Bear in mind that the white side will be visible when the card is complete. This should help you decide which are hill folds and which are valley folds. Check that at each generation the pop-up feature folds entirely flat.

4. When the fractal cut is completed, spread glue on the shaded areas at the back and glue them to the inside of the coloured cover, smoothing outwards from the central fold.

◆

Trisections

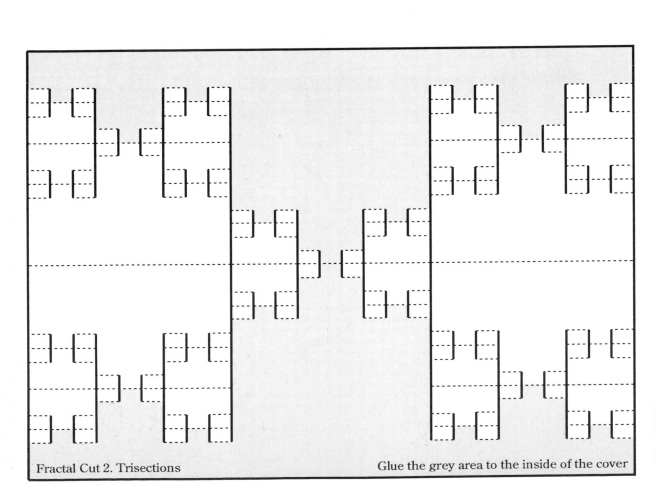

Fractal Cut 2. Trisections

Glue the grey area to the inside of the cover

Further ideas

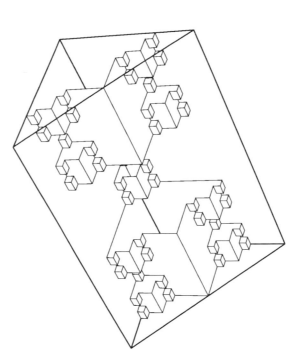

We have already met a strange property of some two-dimensional fractal objects, namely, that an infinite perimeter can enclose a finite area or, indeed, no area at all.

This fractal cut demonstrates the three-dimensional equivalent of this property. Each time we create a new generation we cut along the thicker lines to produce new edges and fold new rectangles to form additional cuboids. In other words, every time we add a new generation, both the total length of the edges and the total volume enclosed by the cuboids increase. However the edges and volume grow in strikingly different ways. As new edges are cut, their total length appears to increase without limit whereas it seems that the volume can never exceed that of a right-triangular prism built upon the first fold.

Since common sense is a rather unreliable guide in matters of convergence or divergence, it is instructive to examine both these situations with care.

Let us start by considering the total volume enclosed. Since fractals are endlessly repeating, the best way to deal with them is by examining the proportional changes from one generation to the next.

At each new generation there are five times as many cuboids, each with one twenty-seventh of the volume. If we say that the volume of the first generation is 1 unit, then the volume of the object after the second generation is

$$1 + \frac{5}{27}$$

After an infinity of generations, the total volume is

$$1 + \frac{5}{27} + \left(\frac{5}{27}\right)^2 + \left(\frac{5}{27}\right)^3 + \cdots$$

This is a geometric series with common ratio less than 1 and its limit is $\frac{27}{22}$.

By such an argument we can see that the fractal cut would never occupy all of what might be called the 'common sense' right-triangular pyramid with volume 3 units.

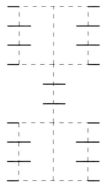

This illustration shows the cuts which are added between generations one and two. Taking the length of the shortest cut to be a single unit, we can see that the total of new lengths being added amounts to 28 units. Using the same measure the total length of cut in the first generation is 16 units. This process increases it in the ratio of 28:16 or 7:4 at each generation. The total length generates this series

$$1 + \frac{7}{4} + \left(\frac{7}{4}\right)^2 + \left(\frac{7}{4}\right)^3 + \cdots$$

This is also a geometric series but its common ratio is greater than 1 and therefore it is divergent. The length of cut edge increases without limit.

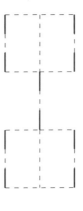

A feature of this fractal cut is that all the development takes place along valley folds. Any hill fold which is created remains unchanged in future generations. It might seem therefore that since valley folds are being converted into hill folds generation by generation, in the limit all folds would become hill folds.

We should then have the paradox of a pop-up card which is able to protrude into three-dimensional space and yet have no valley folds which would allow it to do so! We can now examine this paradox by looking at the lengths of valley and hill folds generation by generation.

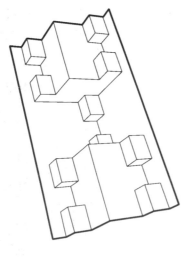

If we call each marked length of valley fold in the diagram above one unit, we can see that the total length is 20 units. In the previous generation there were five lengths of valley fold each three of these units long.

Creating a new generation increases the total length of valley fold in the ratio of 20:15 or 4:3. Because this ratio is greater than 1, the total length of valley fold increases with each new generation and in the limit becomes infinitely long, just as the lengths individually become infinitely small.

In a similar way, the new generation adds ten lengths of hill fold to the two lengths of three units in the previous generation. Creating a new generation therefore increases the length of hill fold in the ratio 10:6 or 5:3.

The total length of hill fold is therefore

$$\frac{2}{3}\left(1 + \frac{5}{3} + \left(\frac{5}{3}\right)^2 + \left(\frac{5}{3}\right)^3 + \cdots\right)$$

Since the common ratio is greater than 1 the length increases without limit.

Although it is difficult to imagine this fractal cut in the limit, it is the appearance of these two infinities which generates the paradox. It is yet another indication of the strangeness of fractal objects. The total length of valley fold reduces to zero, but the number of valley folds becomes infinite.

This paradoxical situation has a curious connection with the paradox proposed by the Ancient Greek philosopher Zeno. He used it to deny the possibility of movement. Zeno said that in order to travel between two points, say Athens and Crete, one must first reach the point midway between the two. But in order to reach that point one must first reach the point midway between Athens and that midpoint. This process continues for ever and so the journey can never start! In Zeno's paradox to travel from one point to another one must travel through an infinity of segments each infinitely small. In this fractal cut we must fold an infinity of segments, each infinitely short.

From a mathematical point of view Zeno's paradox was solved around the turn of the century by the German mathematician Cantor with his theory of transfinite numbers, thus paving the way for an understanding of fractal objects more than half a century later.

FRACTAL CUTS

Card 3. *Alternating Steps*

◆

An interesting feature of this fractal cut is that each generation is the mirror image of the previous one.

A casual glance at the sizes of the cuboids on this card might suggest that there are four generations present. In fact, there are only two, one a mirror reflection of the other.

With a scale factor of a quarter, the sizes reduce rapidly and a third generation would be too small to cut and fold easily.

To make a card with three generations would mean starting with a much larger piece of paper.

Alternating Steps

The initial fold

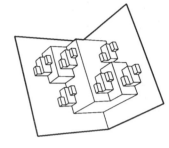

The first generation

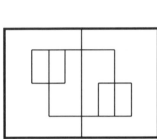

After two generations

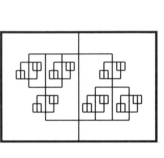

and so on...

From fractal to fractal cut

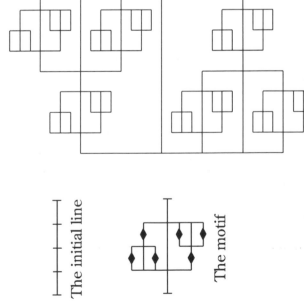

The initial line

The motif

This diagram shows the two generations of the fractal which have been used to make the pop-up card, together with the initial line and the motif which generates it.

The initial line is divided in the ratio 1:2:1 and the motif developed on it from a square and two half-size squares.

The motif has six active lines and the new generation is produced by replacing each of the active lines with a complete motif, one quarter of its size and the mirror image of the previous generation.

| Scale Factor 1:4 | Multiplication Factor 6:1 |

To convert this geometrical figure into a fractal cut, all lines at right angles to the initial line become cuts. All the other lines are scored and folded.

Work generation by generation and use the diagrams on the right to determine which folds are hill folds and which are valley folds in the finished card.

How to make
POP-UP CARD 3

This card consists of two parts, the fractal cut on this page and the cover on page 81.

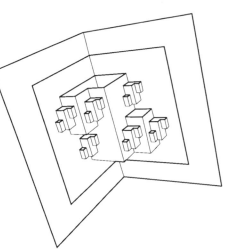

1. Remove both pages from the book and then cut out the pieces precisely.

2. Score along the fold lines and then cut along the cut lines with a scalpel.

-------------- score line

_____ cut line

3. Fold and crease the fractal cut, one generation at a time, using the diagrams opposite for guidance. Bear in mind that the white side will be visible when the card is complete. This should help you decide which are hill folds and which are valley folds. Check that at each generation the pop-up feature folds entirely flat.

4. When the fractal cut is completed, spread glue on the shaded areas at the back and glue them to the inside of the coloured cover, smoothing outwards from the central fold.

Alternating Steps

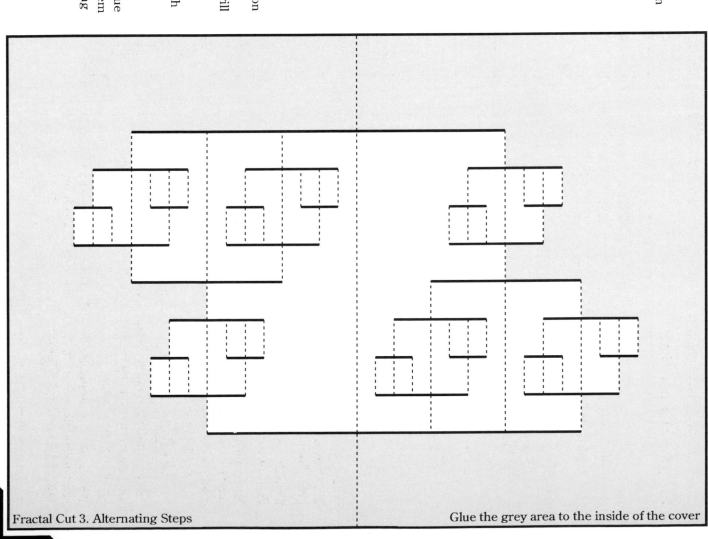

Fractal Cut 3. Alternating Steps

Glue the grey area to the inside of the cover

Further ideas

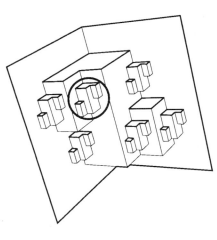

Such an alternating procedure does not destroy the fractal nature of the design. However to reproduce it exactly by enlargement from any particular generation does require that we know whether it is an odd or an even generation. Alternatively we need to know the handedness of the first generation.

This illustration shows a well known fractal, usually called the H-fractal. It makes a very attractive design and is a relatively simple one to reproduce accurately by drawing. It is also a good one to start with when learning to draw fractals on a computer. The motif is symmetrical so the notion of mirror images does not arise.

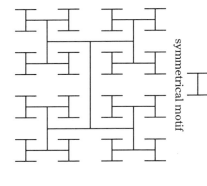

symmetrical motif

Let us now examine what happens when the right upright is halved but otherwise the H-fractal motif is left unchanged. This small modification introduces asymmetry and the mirror image is different. Since each generation is to be a mirror image of the previous one, the fractal develops in a rather different way.

motif   mirror image

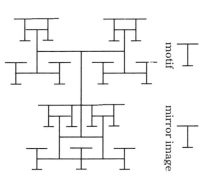

Of course one would expect it to develop in a different way, but one curious change is how some elements begin to coincide. For instance, with a multiplication factor of four, one would expect there to be 16 third-generation elements. In fact there are only 15. Can you identify which element is really of double thickness?

Since fractals are self-similar, we should expect that if we enlarge the circled area by a factor of four, we should reproduce the whole object. Granted that the fractal cut shows just two generations and that it really has infinite detail, this is undoubtedly true. Except for the fact that we seem to have acquired a mirror image!

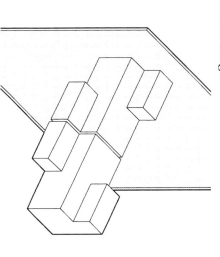

The generating rule for this fractal cut includes the proviso that each generation is the mirror image of the previous one. Hence as the generations continue, all the odd-numbered generations have the orientation of the motif in the illustration above, and all the even-numbered generations have the orientation of its mirror image.

Fractal Cut 3

As we have just seen, the use of the mirror image of an asymmetrical motif introduces striking and perhaps unpredictable results. For instance, let us now look again at the motif of the tropical weed, this time together with its mirror image.

Now, instead of replacing every segment by a reduced copy of the motif as we did before, we only replace certain segments, those marked with solid lines. Those marked with dashed lines we replace by mirror images.

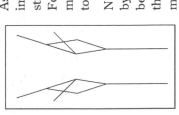

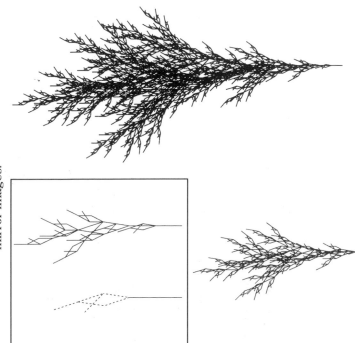

The weed immediately begins to develop in a different way and becomes much more symmetrical. After a few generations the weed is recognisably the same one, but we have stopped the wind from blowing across the tropical plain!

Of course the choice of which segments to replace with mirror images is an arbitrary one and opens up an interesting field for investigation. Different choices can produce very different results and the illustration above shows a plant generated by just such a choice.

FRACTAL CUTS

Card 4. Contrary Cubes

◆

All the cards demonstrate mechanisms which pop up, but this also includes one which pops down.

The central geometrical figure is a pop-up cube with a pop-down cubical recess within it. This motif of contrary cubes grows from the central two quarters of the active line.

Each new generation appears only in valley folds and therefore each cube and its recess is a quarter the size of the previous one.

Working on this scale the card is only conveniently able to show two generations, but it is plain that the fractal design continues for ever.

Contrary Cubes

The initial fold

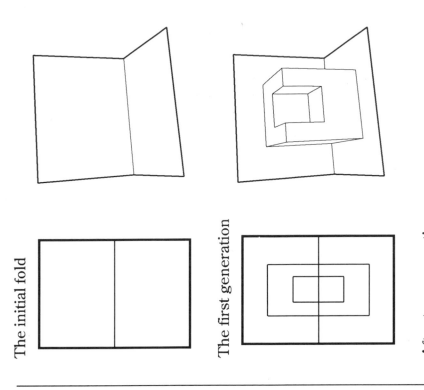

The first generation

After two generations

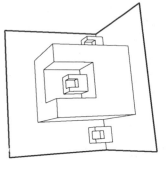

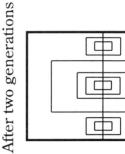

and so on...

From fractal to fractal cut

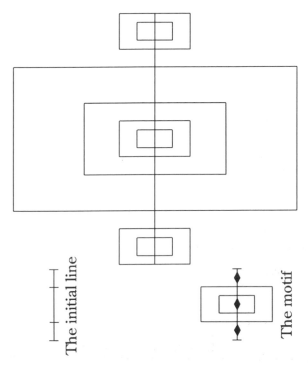

The initial line

The motif

This diagram shows the two generations of the fractal which have been used to make the pop-up card, together with the initial line and the motif which generates it.

The initial line is divided in the ratio 1:2:1 and the motif developed on the central section. The inner rectangle also divides the central section in the ratio 1:2:1.

The motif has three active lines and the next generation is produced by replacing each of the active lines with a complete motif, one quarter its size.

| Scale Factor 1:4 | Multiplication Factor 3:1 |

To convert this geometrical figure into a fractal cut, all lines at right angles to the initial line become cuts. All other lines are scored and folded.

Work generation by generation and use the diagrams on the right to determine which folds are hill folds and which are valley folds in the finished card.

How to make
POP-UP CARD 4

This card consists of two parts, the fractal cut on this page and the cover on page 83.

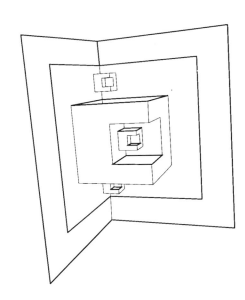

1. Remove both pages from the book and then cut out the pieces precisely.

2. Score along the fold lines and then cut along the cut lines with a scalpel.

-------------------   _____

score line   cut line

3. Fold and crease the fractal cut, one generation at a time, using the diagrams opposite for guidance. Bear in mind that the white side will be visible when the card is complete. This should help you decide which are hill folds and which are valley folds. Check that at each generation the pop-up feature folds entirely flat.

4. When the fractal cut is completed, spread glue on the shaded areas at the back and glue them to the inside of the coloured cover, smoothing outwards from the central fold.

Contrary Cubes

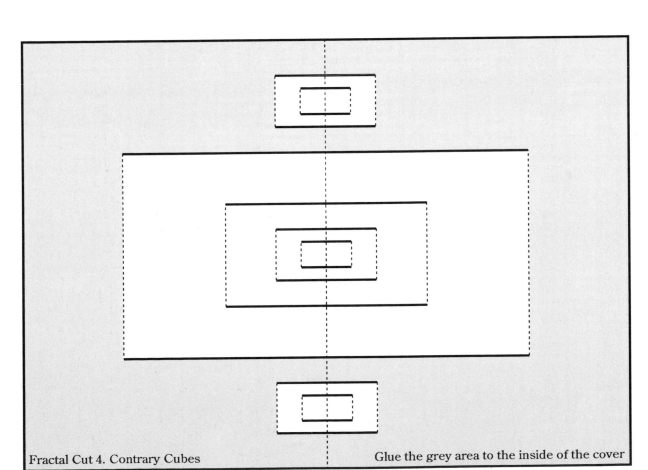

Fractal Cut 4. Contrary Cubes   Glue the grey area to the inside of the cover

Further ideas

If we look sideways at this fractal cut we can see a clearly developing fractal design. Perhaps not one which would have immediately sprung to mind when embarking upon a drawing exercise, but none the less one which makes a very attractive fractal cut.

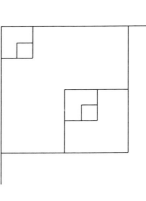

Throughout this discussion of fractals, a great deal of emphasis has been place on the rigidity with which the generating rule has to be applied, generation by generation. However we must not lose sight of the fact that the choice of which generating rule to apply is very much under the control of the designer. In this sense, generating rules are arbitrary. The skill is to choose ones which give pleasing results.

We have already met some computer-generated fractals where the generating rule included random selection at certain critical points. For each fractal cut we had to consider which lines of the motif to make active lines in order to maintain its pop-up qualities and Fractal Cut 3 introduced the idea of motifs being reflected in alternate generations. The transformations of reflection, rotation and translation can be applied to parts of the motif without destroying its fractal qualities. With an infinity of generations the idea of self-similarity is not easily lost. The generating rule itself can be modified to include such transformations.

Remember however, that the transformations must be done in a systematic way and then applied rigorously, even if in some computer-generated fractals the choice is to leave the choice to chance!

This motif is a good one to show how a carefully chosen transformation can help in developing a fractal design. The basic motif is a kind of wave shape made from five longer and six shorter segments. As always, to create the next generation each segment is replaced by a complete motif. The motif must be reduced by the appropriate scale factor for each of the two lengths of segment. The problem is then that some of the lines of the next generation coincide and overlap and the design quickly becomes a mess.

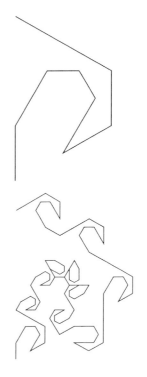

However, if we apply artistic judgment to the choice of whether to reflect or rotate certain portions, we can modify it to get a clearly developing motif which has a continuous line and no ugly overlaps. It is developed further overleaf.

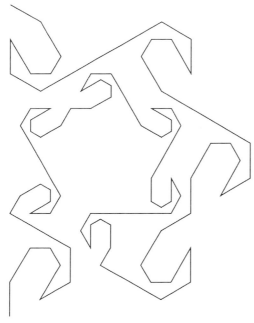

Fractal Cut 4

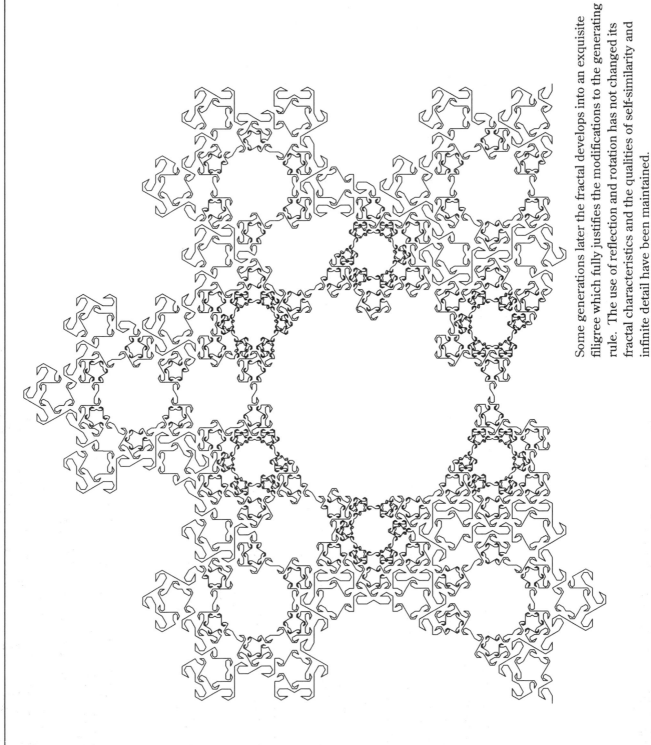

Some generations later the fractal develops into an exquisite filigree which fully justifies the modifications to the generating rule. The use of reflection and rotation has not changed its fractal characteristics and the qualities of self-similarity and infinite detail have been maintained.

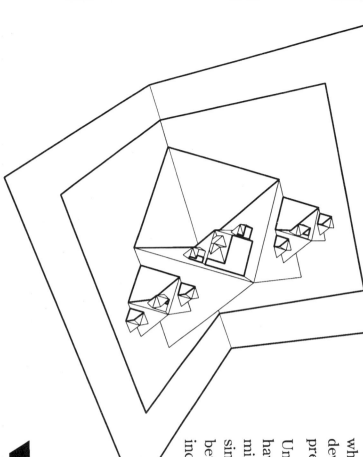

FRACTAL CUTS

Card 5. Tetrahedron Pairs

◆

Cuts at 60° to the original fold line allow the creation of a pair of pop-up tetrahedra which meet at a point.

Each tetrahedron is represented by just two faces which are of course equilateral triangles and all new development takes place along the valley folds of the previous generation.

Unlike the cards based on cuboids this one does not have side to side symmetry. Although this drawing might suggest that the design is symmetrical, with a similar development taking place along valley folds between hidden faces, it is not. A design which included them would not pop up satisfactorily.

Tetrahedron Pairs

The initial fold

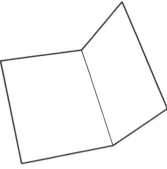

The first generation

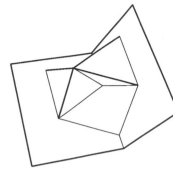

After two generations

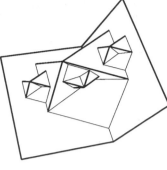

and so on...

From fractal to fractal cut

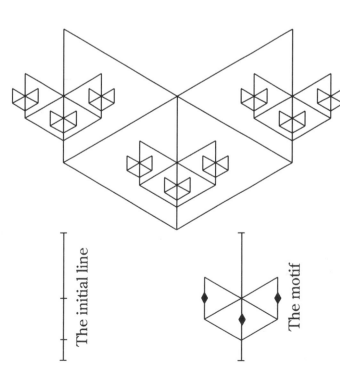

The initial line

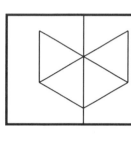

The motif

This diagram shows the three generations of the fractal which have been used to make the pop-up card, together with the initial line and the motif which generates it.

The initial line is divided in the ratio 1:2:3 and the motif developed on it with four equilateral triangles.

The motif has three active lines and the next generation is produced by replacing each of the active lines with a complete motif, one third its size.

| Scale Factor 1:3 | Multiplication Factor 3:1 |

To convert this geometrical figure into a fractal cut, four lines of the motif become cuts and five lines are scored and folded.

Work generation by generation and use the diagrams on the right to determine which folds are hill folds and which are valley folds in the finished card.

42

How to make
POP-UP CARD 5

This card consists of two parts, the fractal cut on this page and the cover on page 85.

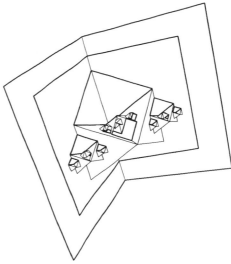

1. Remove both pages from the book and then cut out the pieces precisely.

2. Score along the fold lines and then cut along the cut lines with a scalpel.

- - - - - - - - - score line

——————— cut line

3. Fold and crease the fractal cut, one generation at a time, using the diagrams opposite for guidance. Bear in mind that the white side will be visible when the card is complete. This should help you decide which are hill folds and which are valley folds. Check that at each generation the pop-up feature folds entirely flat.

4. When the fractal cut is completed, spread glue on the shaded areas at the back and glue them to the inside of the coloured cover, matching the A's and B's, smoothing outwards from the central fold.

◆

Tetrahedron Pairs

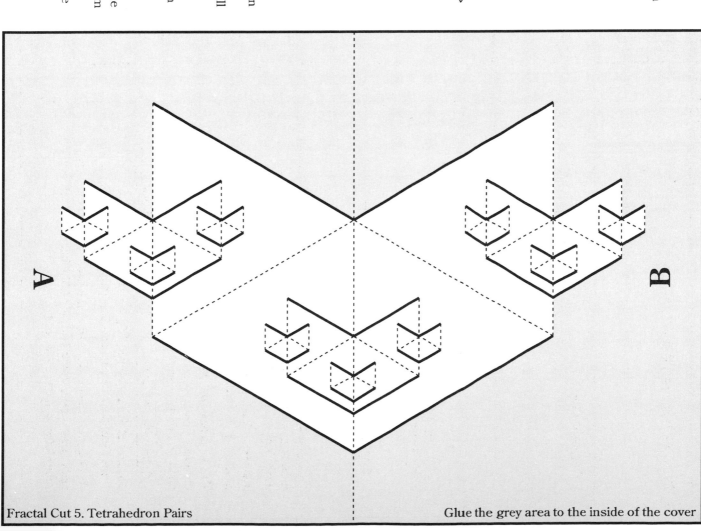

A

B

Fractal Cut 5. Tetrahedron Pairs

Glue the grey area to the inside of the cover

Tetrahedron Pairs

Further ideas

Attractive as this fractal cut is, each tetrahedron is only defined by two faces and the development is not symmetrical side to side. One can now ask if it is possible to construct a pop-up mechanism where each tetrahedron is defined by three faces?

If it is, then the fractal cut would be symmetrical.

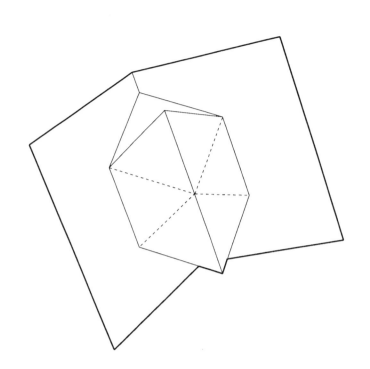

Let us therefore modify the drawing and cutting so that the pop-up part is constructed from a regular hexagon. This will give us six equilateral triangles to fold into two tetrahedra, each defined by three faces. It is clear that a starting configuration is possible.

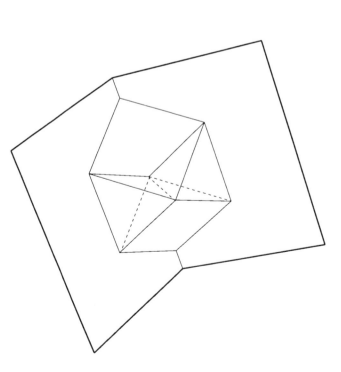

The problem with this new design and the mechanism to achieve it occurs after the card has been opened and when it is being closed again. Just as the tetrahedra are about to become flat, the two valley folds collide. If there is only a single generation a little help is all that is needed to ease them past each other. But in a card with three or four generations this difficulty becomes much greater, making a satisfactory fractal cut impossible.

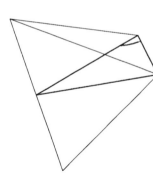

This idea may not in itself be a design for a successful model but it does help us to see that the original 'tetrahedron pairs' fractal becomes a true fractal of tetrahedra only when the card is opened to around 109.5 degrees; not when it is opened to a right angle. Note how the lower end of each valley fold then touches the centre of the base.

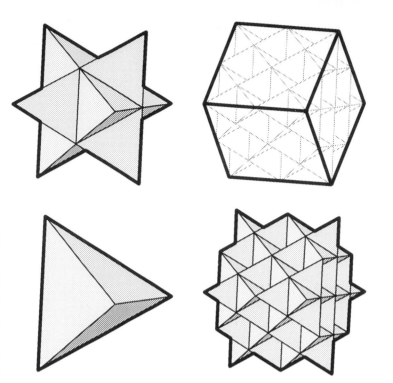

The first generation gives an eight-pointed star which is known as the Stella Octangula. To produce the next generation, each face of this shape must be divided into four triangles. Each triangle becomes a new tetrahedron and so this second generation adds a further 24 vertices to the solid. So it continues, generation by generation. The curious thing about this object is that after an infinite number of stages the exterior of the solid becomes a cube.

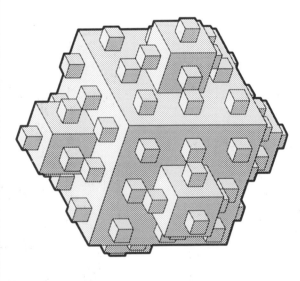

On page 14 we were introduced to a fractal shape made by building smaller cubes upon the faces of the previous generation which was able to fill space without leaving any gaps. This was in spite of a most unpromising start. It is a well-known fact that of all five regular polyhedra (tetrahedron, cube, octahedron, dodecahedron and icosahedron), only cubes are able to fill space without leaving any gaps. Is it true that the shape above is able to fill space only because it is made of cubes?

Is it possible, for instance, to fill space with tetrahedra?

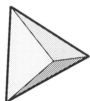

If you need to convince yourself that tetrahedra do not fill space, it is worth making a few models and then trying to fit them together.

Let us now construct a fractal tetrahedron by dividing each face into four triangles and then building a quarter-sized tetrahedron upon the central one.

So a tetrahedron cannot completely fill space, but a fractal based on it can. This is yet another of the strange properties of fractal objects.

FRACTAL CUTS

Card 6. Sierpinski's Triangles

One of the earliest of the two-dimensional fractals to be discovered is known as Sierpinski's Gasket. This card is able to represent it in an interesting pop-up form.

Each triangular depression from one generation is surrounded by the three from the following generation.

The derivation of the pop-up mechanism can also clearly be seen as each triangle is formed by folding opposing diagonals of the two faces of an underlying cuboid.

This fractal design also has strong connections with the pattern of odd and even numbers in Pascal's Triangle.

Sierpinski's Triangles

The initial fold

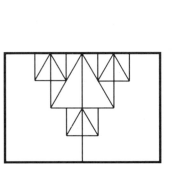

The first generation

After two generations

From fractal to fractal cut

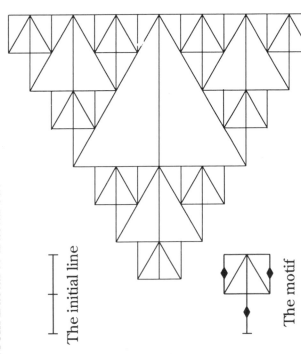

The initial line

The motif

This diagram shows the three generations of the fractal which have been used to make the pop-up card, together with the initial line and the motif which generates it.

The initial line is bisected and the motif developed from two rectangles and their opposed diagonals to create an equilateral triangle.

The motif has three active lines and the next generation is produced by replacing each of the active lines with a complete motif, one half its size.

Scale Factor 1:2   Multiplication Factor 3:1

To convert this geometrical figure into a fractal cut, all the lines of the motif at right angles to the initial line become cuts. All the other lines are scored and folded.

Work generation by generation and use the diagrams on the right to determine which folds are hill folds and which are valley folds in the finished card.

and so on…

How to make

POP-UP CARD 6

This card consists of two parts, the fractal cut on this page and the cover on page 87.

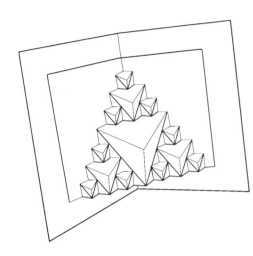

1. Remove both pages from the book and then cut out the pieces precisely.

2. Score along the fold lines and then cut along the cut lines with a scalpel.

- - - - - - - - - -   score line

———————   cut line

3. Fold and crease the fractal cut, one generation at a time, using the diagrams opposite for guidance. Bear in mind that the white side will be visible when the card is complete. This should help you decide which are hill folds and which are valley folds. Check that at each generation the pop-up feature folds entirely flat.

4. When the fractal cut is completed, spread glue on the shaded areas at the back and glue them to the inside of the coloured cover, matching the A's and the B's, smoothing outwards from the central fold.

Sierpinski's Triangles

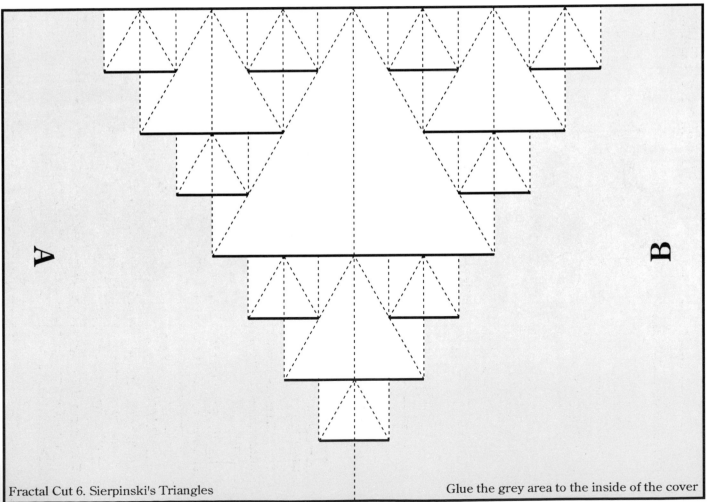

A

B

Fractal Cut 6. Sierpinski's Triangles

Glue the grey area to the inside of the cover

It can be readily seen that this fractal cut is a three-dimensional representation of Sierpinski's Gasket with three generations. This is the same diagram as the one on page 7. There are also interesting connections with Pascal's Triangle.

```
            1
          1   1
        1   2   1
      1   3   3   1
    1   4   6   4   1
  1   5  10  10   5   1
1   6  15  20  15   6   1
1  7  21  35  35  21  7  1
```

Pascal's Triangle is a pattern of numbers which is named after the 17th century French mathematician and philosopher who first wrote a treatise about it. However, it seems to have been known in China and Persia before then, possibly as early as the 10th century A.D.

The pattern is a very simple one, starting with a single digit 1. Each of the other numbers is then calculated by adding the two numbers directly above it. Any empty spaces are regarded as zeros.

Part of the magic of Pascal's Triangle is that it grows by the repeated application of a simple rule from a single 1 in a universe of zeros. The parallel with fractals and their generating rules is strong and it is therefore not surprising that there are connections between them.

The first connection between our fractal cut and Pascal's Triangle is with the pattern of odd and even numbers. Each number is represented by a circle and the odd numbers have been filled with black.

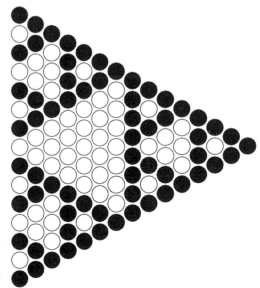

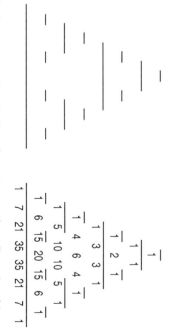

```
            1
          1   1
        1   2   1
      1   3   3   1
    1   4   6   4   1
  1   5  10  10   5   1
1   6  15  20  15   6   1
1  7  21  35  35  21  7  1
```

A further connection is that the pattern of cut lines for this fractal cut corresponds to that obtained by drawing a line over all the odd numbers in the first eight rows of Pascal's Triangle.

Fractal Cut 6

Overleaf we saw that the numbers could be enclosed in circles which were then filled with black or white according to whether they were even or odd.

This method of presentation can also be extended to show whether or not each number is a multiple of some other chosen constant. The illustration below shows the fractal produced when leaving in white the multiples of 12.

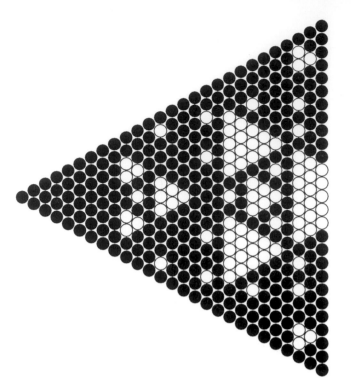

Pascal's Triangle continues for ever, so all such patterns develop more and more detail as they become larger.

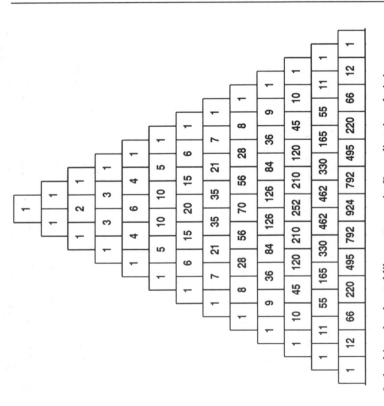

In looking for fractal-like patterns in Pascal's triangle it is beneficial to enclose each of the numbers within a suitable geometric figure. Multiples of different numbers can then be coloured up to a well-defined border and the developing patterns clearly seen. One obvious figure to use is the pattern of squares above.

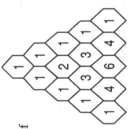

Hexagons make another convenient outline. Each entry is then the sum of the numbers in the two touching hexagons in the row above.

FRACTAL CUTS

Card 7. Fanned Triangles

◇

The triangles for each new generation are rotated clockwise and anticlockwise by 30° about the axis of the previous generation.

After cutting and folding it has to be glued to the base in a non-parallel way in order to create its pop-up properties. Once this is done, the result is a fractal design which folds flat and in which successive generations gradually fan outwards and invade the valleys between earlier generations.

As it develops, several different spirals are generated, creating a complex and beautiful design.

Fanned Triangles

The initial fold

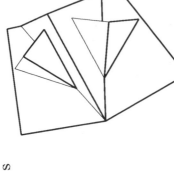

The first generation

After two generations

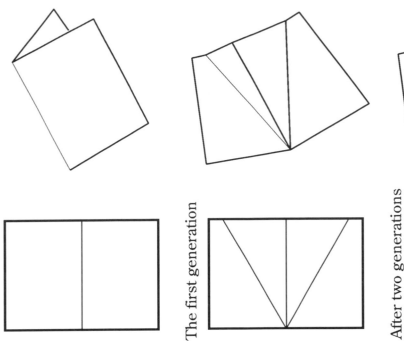

and so on...

From fractal to fractal cut

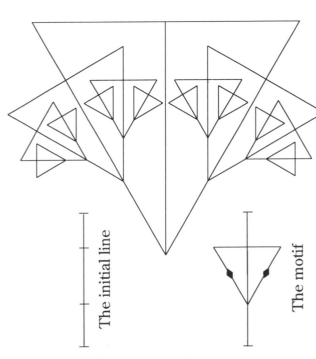

The initial line

The motif

This diagram shows the four generations of the fractal which have been used to make the pop-up card, together with the initial line and the motif which generates it.

The initial line is divided in the ratio 3:4:2 and the motif developed on it with an equilateral triangle.

The motif has two active lines and the next generation is produced by replacing each of the active lines with a complete motif, one half its size.

Scale Factor 1:2   Multiplication Factor 2:1

To convert this geometrical figure into a fractal cut, the line at right angles to the initial line becomes a cut. All the other lines are scored and folded.

Each triangular wedge has a hill fold along its centre and valley folds at its base except where they are interrupted by the smaller wedge of the next generation.

How to make
POP-UP CARD 7

This card consists of two parts, the fractal cut on this page and the cover on page 89.

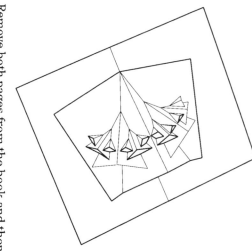

1. Remove both pages from the book and then cut out the pieces precisely.

2. Score along the fold lines and then cut along the cut lines with a scalpel.

- - - - - - - - - - - - score line

———————— cut line

3. Fold and crease the fractal cut, one generation at a time, using the diagrams opposite for guidance. Bear in mind that the white side will be visible when the card is complete. This should help you decide which are hill folds and which are valley folds. Check that at each generation the pop-up feature folds entirely flat.

4. When the fractal cut is completed, spread glue on the shaded areas at the back and glue them to the inside of the coloured cover, matching the A's and the B's.

Fanned Triangles

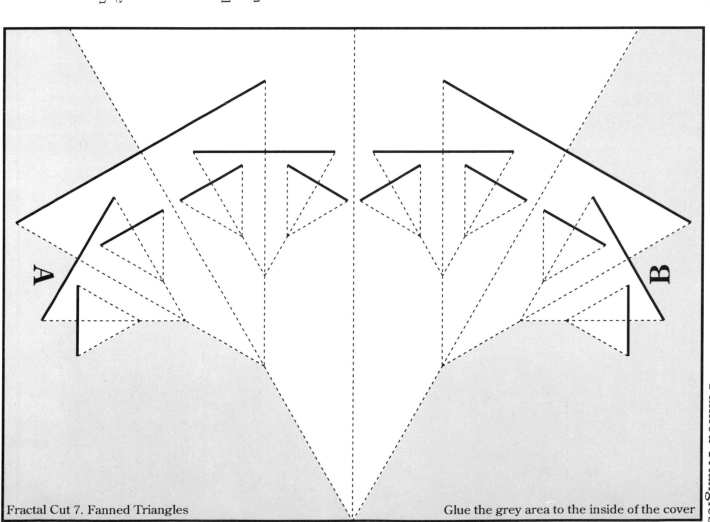

A

B

Fractal Cut 7. Fanned Triangles

Glue the grey area to the inside of the cover

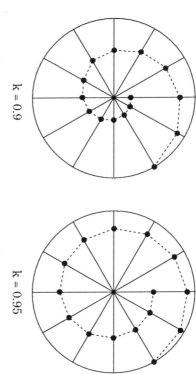

k = 0.9

k = 0.95

Each of these diagrams shows a circle divided into 12 equal sectors. The marked point starts on the circumference and then moves inwards on successive radii. At each step, its distance from the centre is a fixed proportion, k, of its distance on the previous radius. The point traces out a spiral known as the equiangular or logarithmic spiral. Its exact shape depends on the value of k, but it has the remarkable property that it always intersects each radius at the same angle.

It was first discovered by Descartes in 1638 and studied further by Bernouilli. Bernouilli was so taken with the notion that this curve could be reconstituted from a part of itself that he left instructions for it to be carved on his tomb, together with the phrase, *'Eadem mutata resurgo'* which is translated as, 'I shall arise the same though changed'.

Equiangular spirals occur in nature, as in this nautilus shell, probably because this spiral allows a creature to keep its shape as it grows.

The equiangular spiral seems to offer an echo of fractal ideas: it continues for ever, obeying an essentially simple rule and it is self-similar in certain respects. But in another sense it is also quite different since it is a continuous curve and therefore the idea of successive generations cannot apply. Nor could one make a pop-up model of it!

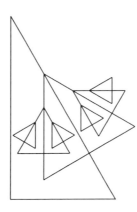

Let us now look at half of the motif which generates the fractal cut. At each generation the triangular motifs are rotated through 30° and each is a fixed proportion of the size of the previous generation. It might therefore seem that perhaps it could generate an equiangular spiral. There are spirals to be found, but since the centre of rotation changes with each generation, they are not true equiangular spirals.

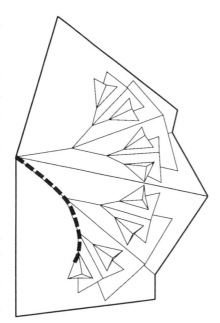

In order to make the fractal cut pop up at all, the edges have to be glued non-parallel to the central fold. When this is done the tips of successive generations of wedges trace out one of these spirals.

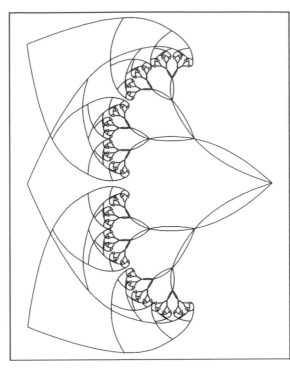

As the generations continue, the wedges become ever smaller and in the limit, when the additional wedges become infinitely small, the three spirals meet at a point. Of course there are other spirals which can also be identified and if we draw all those linking sets of corresponding vertices we find the whole fractal cut is full of crossing spiral paths. These paths in their turn generate a further wonder, a spiral fractal.

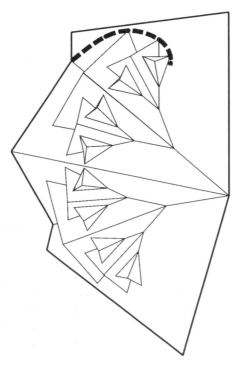

Those vertices of the wedges which lie on the base lie on a second spiral.

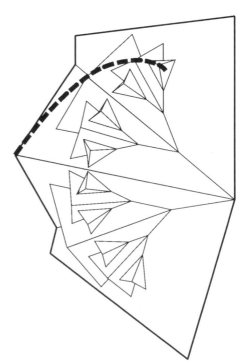

And there is yet a third spiral which can be traced, linking the apexes of the wedges.

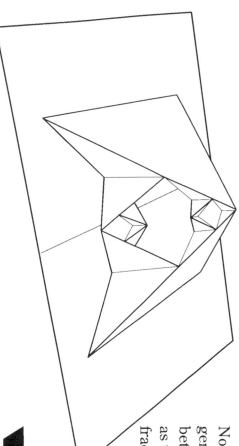

FRACTAL CUTS

Card 8. Opposed Pyramids

This design develops along the valley folds of a pop-up square pyramid.

From each valley two new pyramids are created, each of which is one third the size of the previous generation and which point in opposite directions. Successive generations are rotated by 90° so that all odd numbered generations point vertically upwards and downwards and all even numbered generations point horizontally.

Notice how the remaining vertex of the previous generation can also be regarded as a third pyramid between the opposing pair. It is always the same size as the pair of opposed pyramids which give this fractal cut its name.

Opposed Pyramids

The initial fold

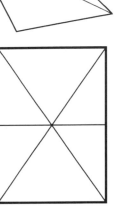

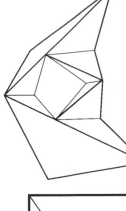

The first generation

After two generations

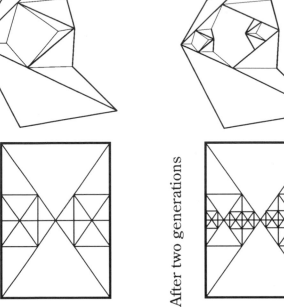

and so on...

From fractal to fractal cut

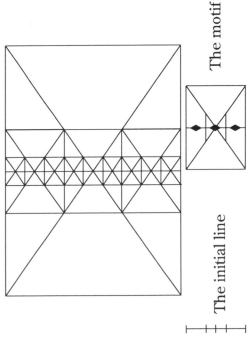

The motif

The initial line

This diagram shows the three generations of the fractal which have been used to make the pop-up card, together with the initial line and the motif which generates it.

The initial line is divided in the ratio 2:1:1:2 and the motif developed on it. The motif consists of a rectangle, with its sides in the ratio √2:1, and its diagonals. There are also two lines parallel to the longer side and one third its length.

The motif has three active lines and the next generation is produced by replacing each of the active lines with a complete motif, one third its size.

Scale Factor 1:3   Multiplication Factor 2:1

To convert this geometrical figure into a fractal cut, lines of the motif which are at right angles to the initial line and which are not already cut, are cut. Some other lines are scored and folded.

This fractal cut is different from all the others because some lines of the diagram are not used to make the pop-up card. Two triangular faces of the pyramid at each stage are left untouched. All development takes place along those faces of the pyramids which have folds on them.

How to make
POP-UP CARD 8

This card consists of two parts, the fractal cut on this page and the cover on page 91.

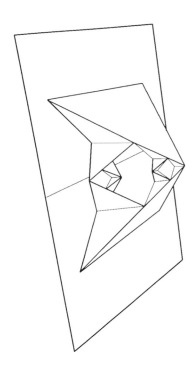

1. Remove both pages from the book and then cut out the pieces precisely.

2. Score along the fold lines and then cut along the cut lines with a scalpel.

‑‑‑‑‑‑‑‑‑‑‑‑‑
score line

───────
cut line

3. Fold and crease the fractal cut, one generation at a time, using the diagrams opposite for guidance. Bear in mind that the white side with no printing on it will be uppermost when the card is complete. This should help you decide which are hill folds and which are valley folds. Check that at each generation the pop-up feature folds entirely flat.

4. When the fractal cut is completed, spread glue on the appropriate sides on the two flaps and glue them to the inside of the coloured cover.

Opposed Pyramids

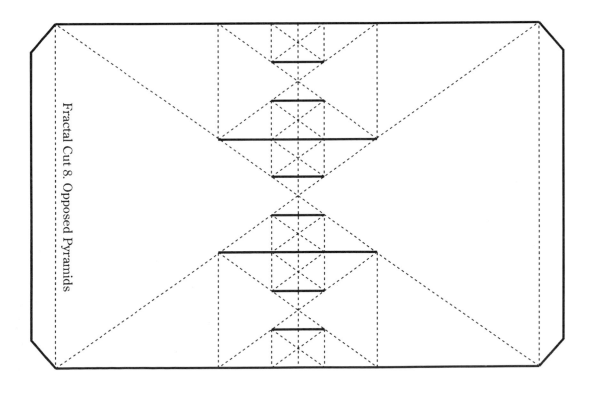

Fractal Cut 8. Opposed Pyramids

When this card is fully opened, the initial pyramid stands upon a square base and its height is half the length of the base. The valley folds along which the fractal development takes place then stand vertically and touch each other. Subsequent generations are one third of the size but of course maintain the same proportions.

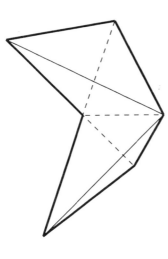

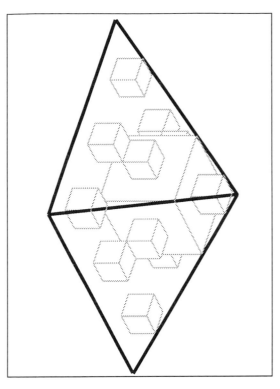

We have already met this pyramid when considering the three-dimensional fractal based upon a cube with one-third sized cubes attached to the centre of the faces of the previous generation. Its volume is one twelfth of the complete rhombic dodecahedron.

A pyramid which has these proportions placed inside a cube will have its vertex at the centre of the cube.

The centre of the cube is the meeting point of six of these pyramids and it can be seen that six of these pyramids glued on the outside of a cube will give the rhombic dodecahedron.

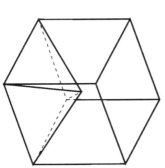

Without some models to handle, it may not be so easy to be certain that rhombic dodecahedra do in fact fill space without any gaps. However, thinking about this pyramid and its properties is very helpful. Let us imagine a network of cubes which do fill space. Then divide alternate cubes in every direction into six of these pyramids. All we then have to do is to separate them mentally and to imagine each to be attached to a face of the neighbouring uncut cube. Such a process does not introduce any gaps and so we can be certain that rhombic dodecahedra do pack together and do fill space.

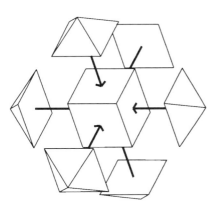

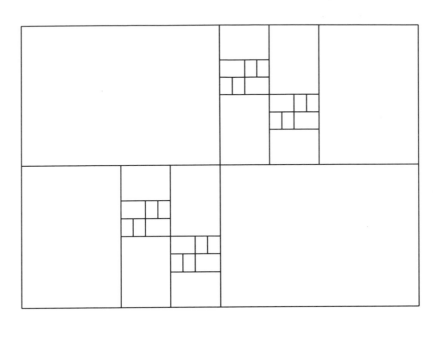

The properties of this fractal cut and its pyramids depend on starting with a rectangle in the proportions of 1:√2.

This same proportion is widely used as a standard for international paper sizes. Why was it selected as the ideal proportion for such a standard? Not because it is a specially pleasing shape. It is generally considered that the famous golden rectangle, whose sides are in the proportion of 1:1.618 is more beautiful.

The Ancient Greeks used the proportions of the golden rectangle in building the Parthenon and other temples and it was also used by Leonardo da Vinci and other Italian artists of the Renaissance to define the basic structures of their works. Even in our own century, Le Corbusier, the famous French architect, developed a system of proportions for use in buildings based on the golden proportion.

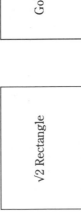

√2 Rectangle   Golden Rectangle

The reasons for the use of the 1:√2 rectangle is rather more technical and less mystical. If you take a sheet of paper which has sides in this proportion and halve it, the sides of the half sheets are also in the proportion of 1:√2. Likewise, if you place two such sheets side by side along their longer edges, the larger sheet is also in the proportion of 1:√2.

The 'A' system of paper sizes, using this proportion, is therefore a valuable way of eliminating waste when cutting paper into smaller pieces. The largest size is A0 which has an area of one square metre. Smaller sizes are known as A1, A2, A3, A4, A5, etc. Each 'generation' has half the area of the previous one.

All this sounds familiar to us. We are considering a sheet made of similar smaller sheets, made in turn of ever smaller ones in the same proportions. It looks like a straight path to self-similarity and so we shall use it to produce a simple fractal.

Start with a rectangle with its sides in the proportions of 1:√2. Divide it into quarters. Select a pair of opposite quarters and divide them into halves by joining the mid-points of the longer sides, leaving the other two quarters untouched. Continue the process of division, stage by stage, alternating between quartering and halving.

Notice that all the rectangles of this fractal, whatever their size or orientation, are similar. This is only possible if we start with a rectangle with sides in the proportion of 1:√2.

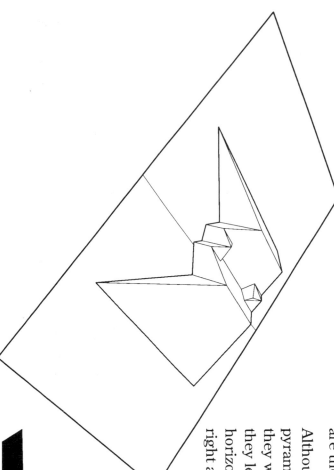

FRACTAL CUTS

Card 9. Disappearing Vertices

◇

A process of dividing and folding the vertex of a pyramid modifies it to create a hexagonal valley with a one-third sized pyramid at each end. All the pyramid vertices which are created in one generation are therefore destroyed to make the next generation.

Although doubling in number at each generation, the pyramids become smaller and smaller and in the limit they will disappear altogether. As they disappear, they leave behind an infinite sequence of alternate horizontal and vertical valley folds lying in a plane at right angles to the opened base of the fractal cut.

Disappearing Vertices

The initial fold

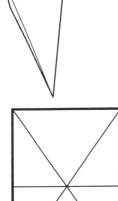

The first generation

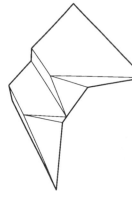

After two generations

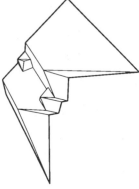

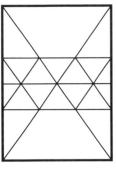

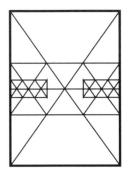

and so on...

From fractal to fractal cut

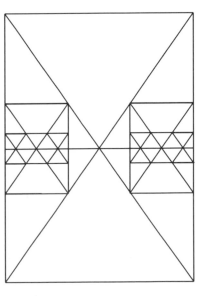

 The initial line

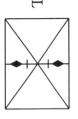 The motif

This diagram shows the two generations of the fractal which have been used to make the pop-up card, together with the initial line and the motif which generates it.

The initial line is divided in the ratio 2:1:1:2 and the motif developed on it. The motif consists of a rectangle with its sides in the ratio √2:1 and its diagonals.

The motif has two active lines, the outer thirds, and the next generation is produced by replacing each of the active lines with a complete motif, one third its size.

Scale Factor 1:3   Multiplication Factor 2:1

To convert this geometrical figure into a fractal cut needs some care. In a sense, the motifs occur in threes and only the outer edges of the sets of three are cut. Since this is the outer edge of the card for the first generation, it is only on the second generation that cuts need to be made.

Work generation by generation. Each pyramid is folded first and then modified to get the hexagonal valley and the next generation of pyramids.

How to make

POP-UP CARD 9

This card consists of two parts, the fractal cut on this page and the cover on page 93.

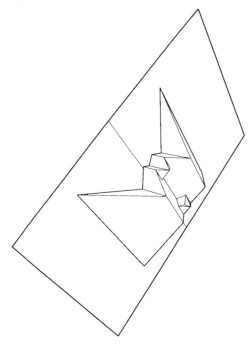

1. Remove both pages from the book and then cut out the pieces precisely.

2. Score along the fold lines and then cut along the cut lines with a scalpel.

- - - - - - - - - - - - - - - -
score line   cut line

——————

3. Fold and crease the fractal cut, one generation at a time, using the diagrams opposite for guidance. Bear in mind that the white side with no printing on it will be uppermost when the card is complete. This should help you decide which are hill folds and which are valley folds. Check that at each generation the pop-up feature folds entirely flat.

4. When the fractal cut is completed, spread glue on the appropriate sides on the two flaps and glue them to the inside of the coloured cover.

Disappearing Vertices

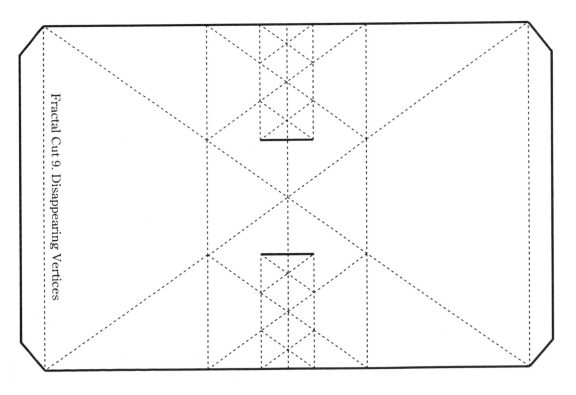

Fractal Cut 9. Disappearing Vertices

Further ideas

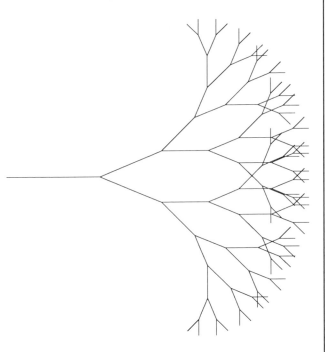

In a certain sense, this illustration returns us to the first fractal we met in this book, the tropical weed. This one is a fractal model of a tree; not a very realistic tree because it is rather too symmetrical, but recognisably a tree. A mathematician would call it a binary tree. The fact that it is a binary tree has nothing at all to do with whether it is symmetrical or not, but with the fact that every branch forks into exactly two new branches.

Let us look at Fractal Cut 9 and think of it in terms of a binary tree. The first generation is a square-based pyramid and that corresponds to the trunk. The generating rule means that we have to fold the vertex of the pyramid in such a way that it gives a hexagonal valley with two smaller pyramids, one at each end. The corresponding process in our tree diagram is that the trunk splits into two branches.

The apexes of the second generation pyramids are replaced by valley folds ending in smaller pyramids. In our tree we replace the tips of the branches by forks from which two smaller branches grow. Both processes can continue for ever although five or six generations will be enough for a real tree.

If we now go back to Fractal Cut 8 and look at it from this botanical point of view, we can see that it can be regarded in two alternative ways.

It can either be seen as a binary tree or as a ternary tree, depending on whether we count the third pyramid between the opposed pair of pyramids as truly a member of the next generation or not.

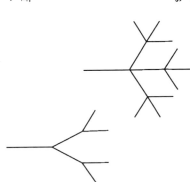

Amongst all possible trees, the binary tree has a special status. A computer, at its fundamental level, consists of millions of minute, unbelievably fast circuits. Each one has only two states, which can be described as on or off, open or shut, or digitally as 1 or 0. Everything that a computer can do has to be seen in terms of the organisation of enormous binary trees of extraordinary complexity.

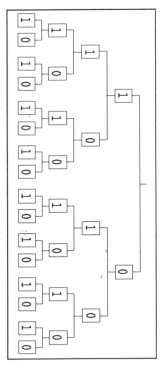

Fractal Cut 9

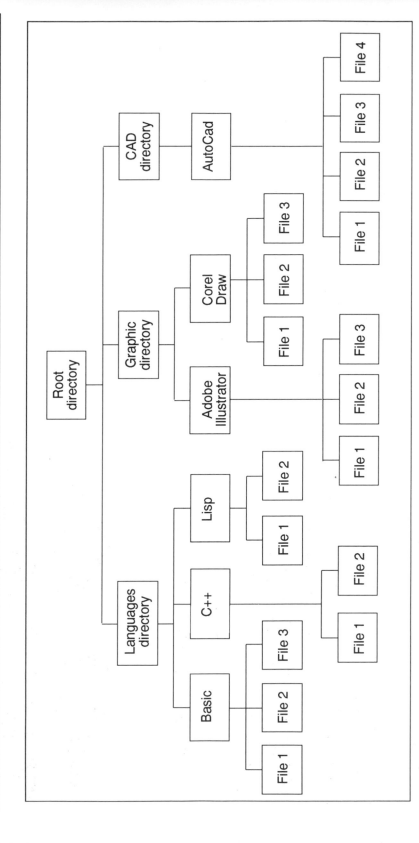

In addition to the use of binary trees at the fundamental level in computers, trees play an important role in information storage and retrieval. The diagram above, for instance, shows how the files are organised and stored in the directory of a typical computer. This structure is seldom simply binary or ternary but has varying numbers of sub-branches sprouting from the ones above. Such trees are not regular fractals since each generation is not identical to the previous one. However, provided they are large enough, they could be regarded as fractals in the statistical sense. Furthermore, a rather amusing difference between a directory tree and our fractal tree is that the trunk is normally placed at the top and is known as the root!

Many programs, such as databases specially designed to store huge amounts of information, also have a tree structure. So the theoretical study of the best procedures to reach the exact place where a certain piece of information is stored in the least possible time has become an important aspect of computer science.

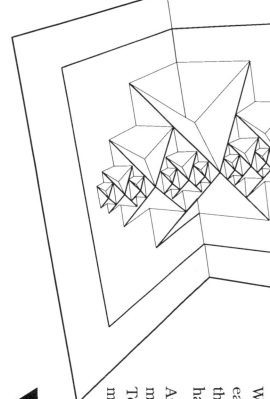

FRACTAL CUTS

Card 10. Hanoi Arête

A progression of diminishing pyramids creates a shape rather like a sharp rising mountain ridge, an arête. Each new generation doubles the number of pyramids and halves their size.

In a way this is reminiscent of the old puzzle about the doubling of grains of corn on the squares of a chess board. The thickness of paper in this fractal cut builds up remarkably.

When the card is closed there are 15 pyramids and each of them, large or small, contributes four thicknesses. After only four generations the cover has to accommodate 60 thicknesses of paper.

Another unexpected property of this card is that it mimics the solution of another famous puzzle 'The Tower of Hanoi' and indicates the order in which the moves have to be made.

Hanoi Arête

The initial fold

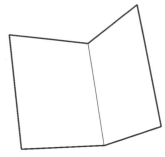

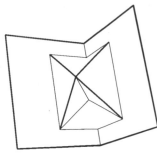

The first generation

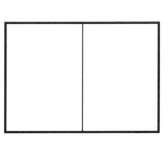

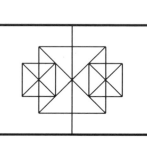

After two generations

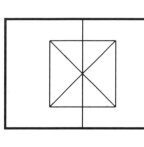

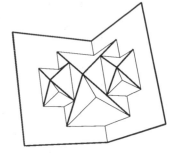

and so on...

From fractal to fractal cut

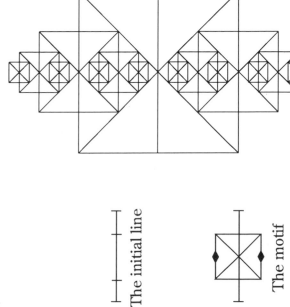

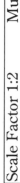

The initial line

The motif

This diagram shows the four generations of the fractal which have been used to make the pop-up card, together with the initial line and the motif which generates it.

The initial line is divided in the ratio 1:2:1 and the motif developed with a square and its diagonals on the central section.

The motif has two active lines and the next generation is produced by replacing each of the active lines with a complete motif, one half its size.

Scale Factor 1:2   Multiplication Factor 2:1

To convert this geometrical figure into a fractal cut, all lines at right angles to the initial line become cuts. All the other lines are scored and folded.

Work generation by generation and use the diagrams on the right to determine which folds are hill folds and which are valley folds in the finished card.

72

How to make
POP-UP CARD 10

This card consists of two parts, the fractal cut on this page and the cover on page 95.

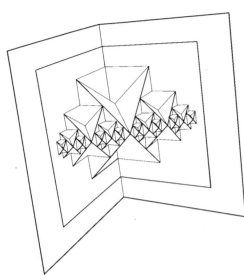

1. Remove both pages from the book and then cut out the pieces precisely.

2. Score along the fold lines and then cut along the cut lines with a scalpel.

------------ score line

———————— cut line

3. Fold and crease the fractal cut, one generation at a time, using the diagrams opposite for guidance. Bear in mind that the white side will be visible when the card is complete. This should help you decide which are hill folds and which are valley folds. Check that at each generation the pop-up feature folds entirely flat.

4. When the fractal cut is completed, spread glue on the grey areas on the back then glue it to the inside of the coloured cover, smoothing outwards from the central fold.

Hanoi Arête

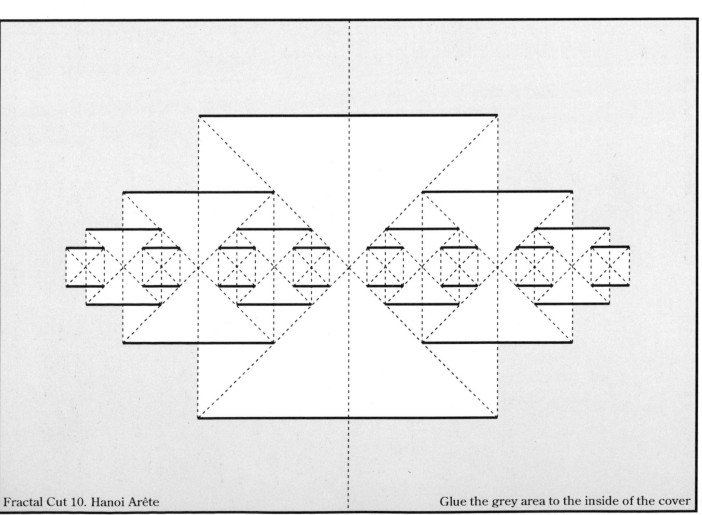

Fractal Cut 10. Hanoi Arête

Glue the grey area to the inside of the cover

Further ideas

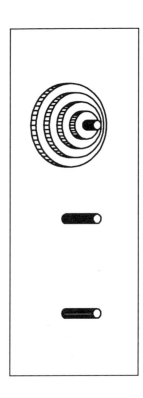

This model illuminates an unexpected relation between fractals and a famous puzzle: the Tower of Hanoi. This puzzle was created in the last century by a French professor of mathematics, Edouard Lucas and has fascinated people ever since.

The puzzle consists of three pegs and a number of disks of different sizes, each with a hole at its centre. To start the puzzle, the disks are stacked on one of the pegs in order of size with the largest at the bottom. The puzzle is to move all the disks to another peg in the least possible number of moves.

The restrictions are that no disk may ever rest on a smaller one and only one disk may be moved at a time.

Mathematicians have shown that the minimum number of moves to effect the transfer is always 2^n-1, where the letter n stands for the total number of disks. Commercially produced Tower of Hanoi puzzles usually have seven disks and therefore it takes 127 moves to complete it. Here we are going to solve it for four disks, so the minimum number of moves must be

$$2^4-1 = 15$$

We know that we can do it in 15 moves but what moves should we make? Our fractal cut provides the answer. Notice that the fractal cut has four generations and we shall match each generation to a disk of a corresponding size.

Place the disks over the left peg. Now trace an imaginary line over the central line of symmetry of the fractal cut, starting at the top.

This line shows you in which order you have to move the disks The line starts at the fourth generation of the fractal, so move the fourth size of disk to the right, to the middle peg. Coming down the line, the next generation we meet is the third, so we must move the third size of disk to the right. It cannot go on top of a smaller disk, so the only possible place for it must be the empty peg. The next generation on the model is again the fourth, so move the fourth size disk to the right and place it on top of the third sized disk.

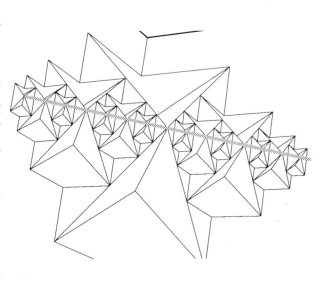

Continue in the same way, each time using the generation on the model to show which disk to move. Always move disks to the right but treat the pegs as if they were arranged around a circle; in other words consider that the leftmost peg also lies to the right of the rightmost! So when in move 5 you need to move the smallest disk, you must place it on the left peg.

After exactly fifteen moves the disks will be placed on the rightmost peg in the correct order and you will have completed the traverse of the fractal cut. Overleaf you will see the solution set out in the form of a diagram.

Fractal Cut 10

Delightful as it is to be able to find a connection like this, you may well have noticed that Fractal Cut 1 could also have been used as the solution. So indeed could the fractal drawing!

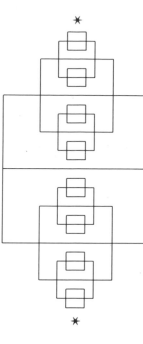

This diagram also has a further curious property. On the axis of the design there are two special points marked with stars. In every generation the centre of the square nearest to each of the marked points is the same distance from it as is the length of the side of that square.

We have often dealt with the lengths, areas and volumes of fractals by summing infinite series. Fractals are the geometrical equivalent of an infinite sequence of terms and so this is not surprising. Many people will be familiar with the sums of infinite geometric series but if not, then the proof is not a difficult one. Let us show the existence of these special points by summing an infinite series directly.

Suppose that the largest square in this fractal has a side of length 2 units. Starting from the centre, we can see that the furthest point reached by the last square of the last generation is given by the sum of an infinite sequence of terms.

$$S = 1 + \frac{1}{2} + \frac{1}{4} + \frac{1}{4} + \frac{1}{8} + \cdots$$

$$2S = 2 + 1 + \frac{1}{2} + \frac{1}{2} + \frac{1}{4} + \cdots$$

Subtracting

$$2S - S = 2 + 0 + 0 + 0 + \cdots$$

$$S = 2$$

Hence, after an infinity of generations, the fractal cut will reach a point which is as far from the centre of the square as the length of the square. This is true whichever square we choose as our starting point.

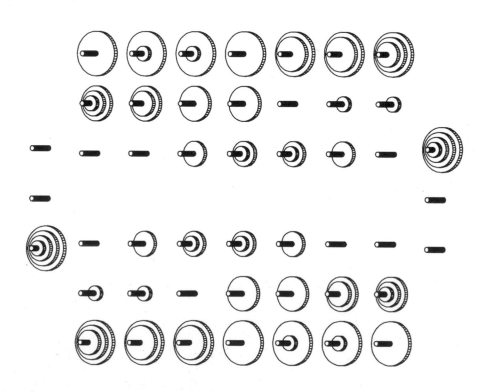

This illustration was created on a computer using a drawing program. Programs like this are able to duplicate images at will and to transform them in many different ways almost instantaneously. A few keystrokes or clicks with a mouse are all that is needed to enlarge or reduce them or to create new composites from earlier images. Using such a program to create an image such as the one above is almost a fractal experience in itself. There is considerable satisfaction in achieving it in a minimum number of operations.

Remove this page from the book and then cut out the cover precisely. Score along the central line and then glue the grey area of the fractal cut from page 19 into the marked position, working outwards from the central fold.

Scale Factor 1:2

CENTRAL QUARTILES

Multiplication Factor 2:1

FRACTAL
CUTS

Central Quartiles

A fractal cut is an endlessly developing pop-up design which belongs to the class of curious geometrical objects called fractals. Starting with a motif and applying a generating rule generation by generation, the design develops more and more detail.

This card only shows four generations, but it can be clearly seen that with infinite patience, infinite skill and infinite time it could be continued for ever.

Apart from their fractal qualities a feature of fractal cuts is that they are made from a single piece of paper with nothing added and nothing taken away.

The Motif
A line divided into quarters with the central two quarters converted into a pop-up cuboid.

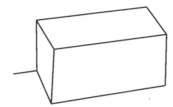

The Generating Rule
Replace each valley fold of the cuboid with a complete motif one half of the size.

Cover for Pop-up Card 2

Remove this page from the book and then cut out the cover precisely. Score along the central line and then glue the grey area of the fractal cut from page 25 into the marked position, working outwards from the central fold.

Scale Factor 1:3   TRISECTIONS   Multiplication Factor 5:1

Trisections

A fractal cut is an endlessly developing pop-up design which belongs to the class of curious geometrical objects called fractals. Starting with a motif and applying a generating rule generation by generation, the design develops more and more detail.

This card only shows three generations, but it can be clearly seen that with infinite patience, infinite skill and infinite time it could be continued for ever.

Apart from their fractal qualities a feature of fractal cuts is that they are made from a single piece of paper with nothing added and nothing taken away.

The Motif

A line divided into thirds with both outer thirds converted into pop-up cuboids.

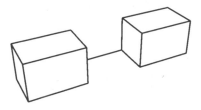

The Generating Rule

Replace each length of valley fold with a complete motif one third of the size.

Remove this page from the book and then cut out the cover precisely. Score along the central line and then glue the grey area of the fractal cut from page 31 into the marked position, working outwards from the central fold.

Scale Factor 1:4

ALTERNATING STEPS

Multiplication Factor 6:1

Alternating Steps

A fractal cut is an endlessly developing pop-up design which belongs to the class of curious geometrical objects called fractals. Starting with a motif and applying a generating rule generation by generation, the design develops more and more detail.

This card only shows two generations, but it can be clearly seen that with infinite patience, infinite skill and infinite time it could be continued for ever.

Apart from their fractal qualities a feature of fractal cuts is that they are made from a single piece of paper with nothing added and nothing taken away.

The Motif
A line divided into quarters with the central two quarters converted into a group of three pop-up cuboids.

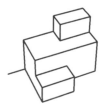

The Generating Rule
Replace each length of valley fold of the cuboids with a complete motif one quarter of the size and the mirror image of the previous generation.

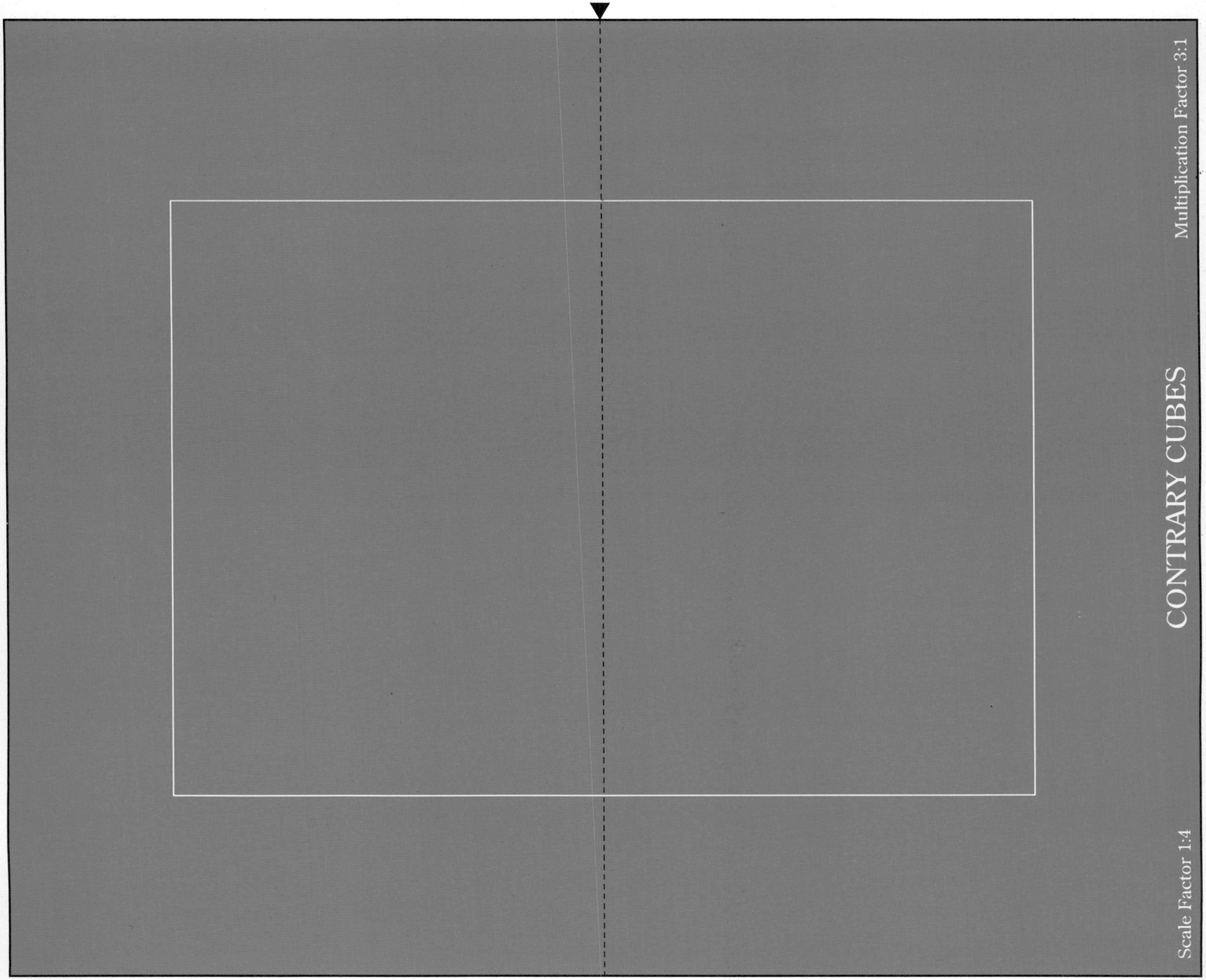

Cover for Pop-up Card 4

Remove this page from the book and then cut out the cover precisely. Score along the central line and then glue the grey area of the fractal cut from page 37 into the marked position, working outwards from the central fold.

Scale Factor 1:4

CONTRARY CUBES

Multiplication Factor 3:1

FRACTAL CUTS

Contrary Cubes

A fractal cut is an endlessly developing pop-up design which belongs to the class of curious geometrical objects called fractals. Starting with a motif and applying a generating rule generation by generation, the design develops more and more detail.

This card only shows two generations, but it can be clearly seen that with infinite patience, infinite skill and infinite time it could be continued for ever.

Apart from their fractal qualities a feature of fractal cuts is that they are made from a single piece of paper with nothing added and nothing taken away.

The Motif
A line divided into quarters with the central two quarters converted into a pop-up cube, which in turn has its two central quarters popped-down.

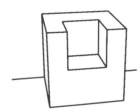

The Generating Rule
Replace each valley fold which is one quarter the length of the original with a complete motif one quarter of the size.

Cover for Pop-up Card 5

Remove this page from the book and then cut out the cover precisely. Score along the central line and then glue the grey area of the fractal cut from page 43 into the marked position, matching A's and B's, working outwards from the central fold.

B

A

Scale Factor 1:3

TETRAHEDRON PAIRS

Multiplication Factor 3:1

Tetrahedron Pairs

A fractal cut is an endlessly developing pop-up design which belongs to the class of curious geometrical objects called fractals. Starting with a motif and applying a generating rule generation by generation, the design develops more and more detail.

This card only shows three generations, but it can be clearly seen that with infinite patience, infinite skill and infinite time it could be continued for ever.

Apart from their fractal qualities a feature of fractal cuts is that they are made from a single piece of paper with nothing added and nothing taken away.

The Motif

A bisected line with a pair of pop-up tetrahedra which centre about the mid point.

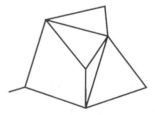

The Generating Rule

Replace each valley fold of the tetrahedron pair with a complete motif one third of the size.

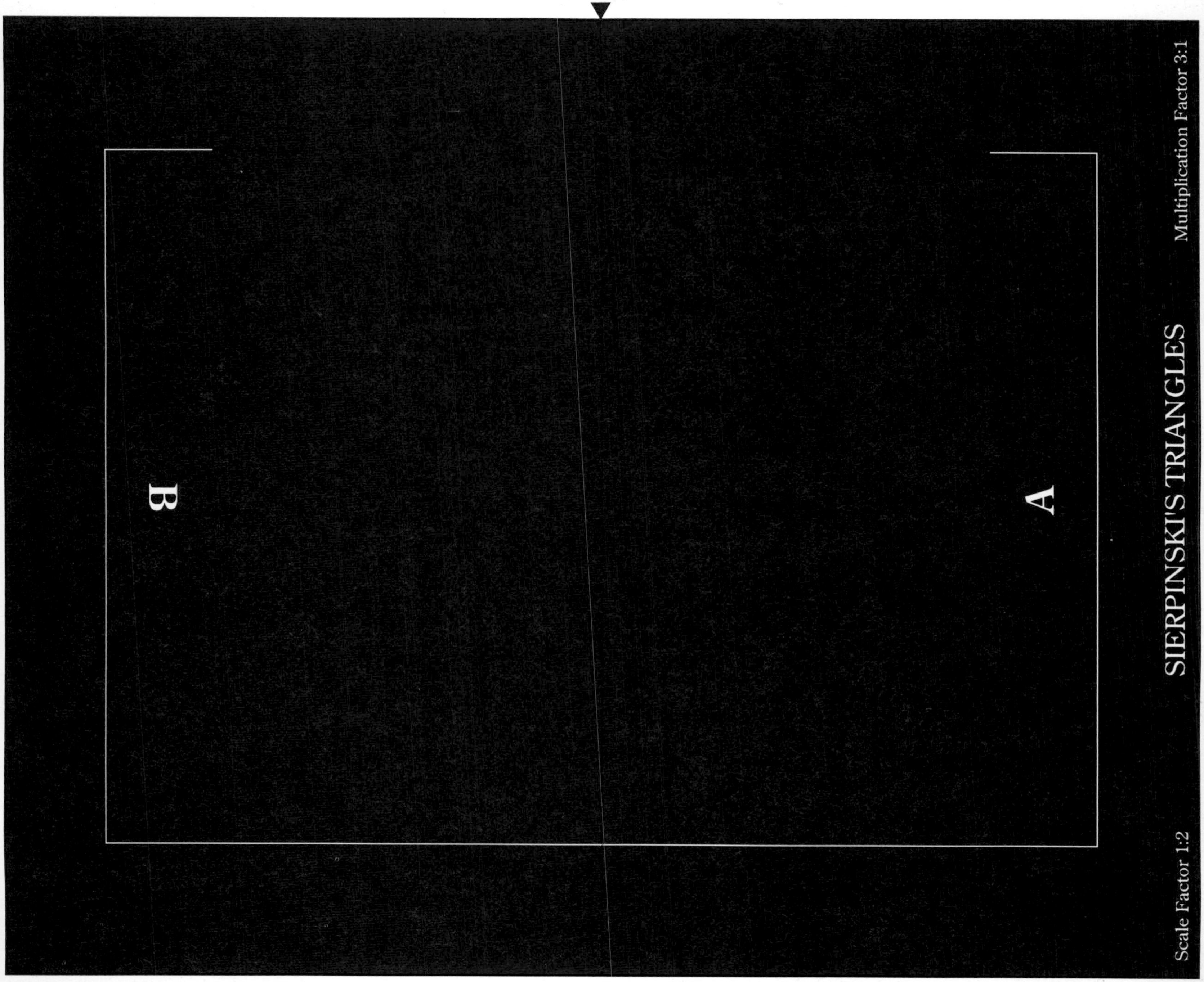

Cover for Pop-up Card 6

B

A

Remove this page from the book and then cut out the cover precisely. Score along the central line and then glue the grey area of the fractal cut from page 49 into the marked position, matching A's and B's, working from the central fold.

Scale Factor 1:2   SIERPINSKI'S TRIANGLES   Multiplication Factor 3:1

Sierpinski's Triangles

A fractal cut is an endlessly developing pop-up design which belongs to the class of curious geometrical objects called fractals. Starting with a motif and applying a generating rule generation by generation, the design develops more and more detail.

This card only shows three generations, but it can be clearly seen that with infinite patience, infinite skill and infinite time it could be continued for ever.

Apart from their fractal qualities a feature of fractal cuts is that they are made from a single piece of paper with nothing added and nothing taken away.

The Motif
A bisected line with a triangle folded from the diagonals of a pop-up cuboid.

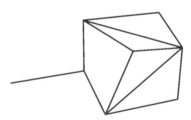

The Generating Rule
Replace each valley fold parallel to the original fold of the card with a complete motif half the size.

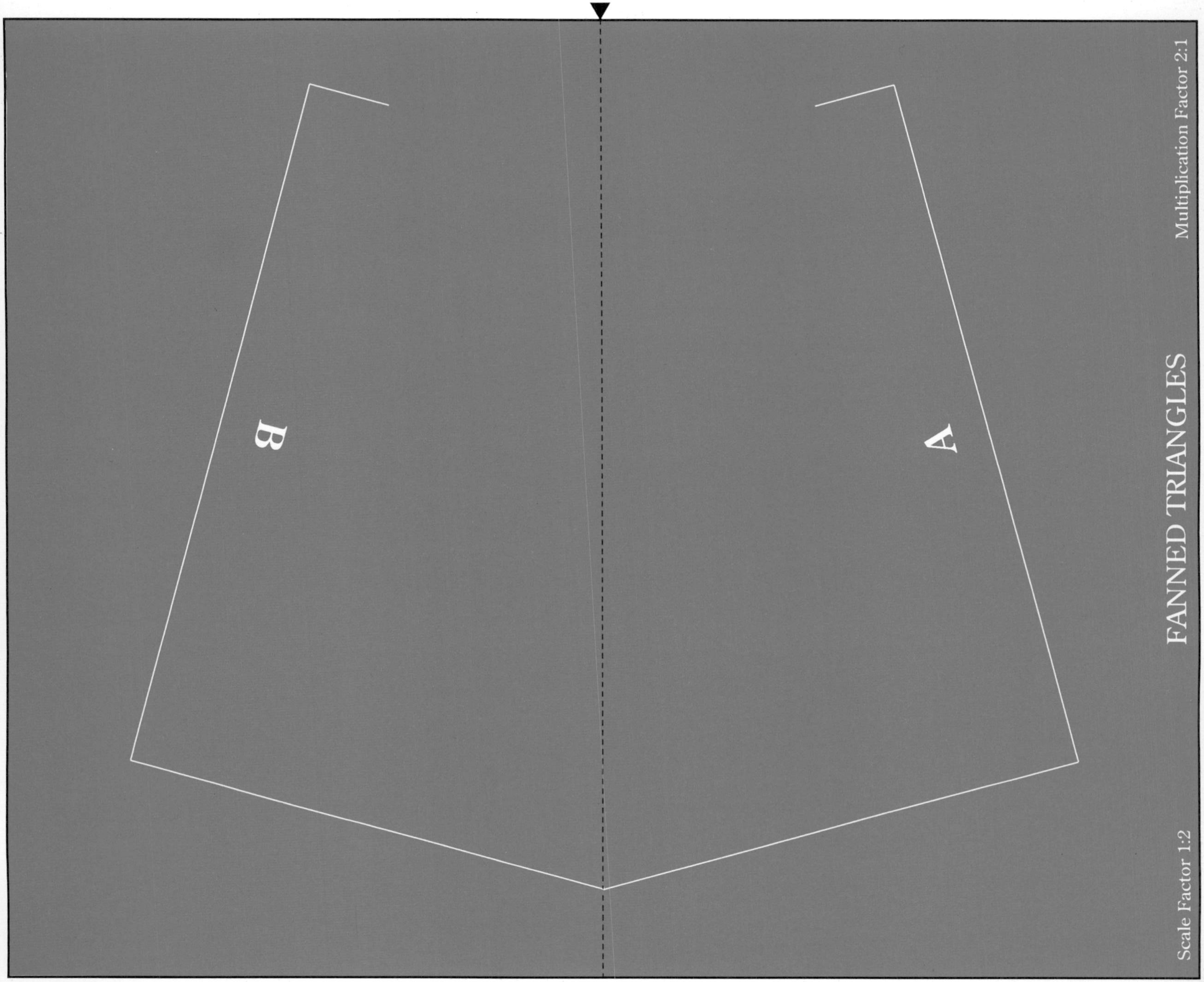

Cover for Pop-up Card 7

Remove this page from the book and then cut out the cover precisely. Score along the central line and then glue the grey area of the fractal cut from page 55 into the marked position, matching A's and B's, working outwards from the central fold.

B

A

Scale Factor 1:2   FANNED TRIANGLES   Multiplication Factor 2:1

FRACTAL CUTS

Fanned Triangles

A fractal cut is an endlessly developing pop-up design which belongs to the class of curious geometrical objects called fractals. Starting with a motif and applying a generating rule generation by generation, the design develops more and more detail.

This card only shows four generations, but it can be clearly seen that with infinite patience, infinite skill and infinite time it could be continued for ever.

Apart from their fractal qualities a feature of fractal cuts is that they are made from a single piece of paper with nothing added and nothing taken away.

The Motif
A line divided into three unequal pieces with the central section converted into a pop-up triangular wedge.

The Generating Rule
Replace each valley fold with a complete motif half the size.

Remove this page from the book and then cut out the cover precisely. Score along the central line and then glue the flaps of the fractal cut from page 61 into the marked positions.

Scale Factor 1:3

OPPOSED PYRAMIDS

Multiplication Factor 2:1

Opposed Pyramids

A fractal cut is an endlessly developing pop-up design which belongs to the class of curious geometrical objects called fractals. Starting with a motif and applying a generating rule generation by generation, the design develops more and more detail.

This card only shows three generations, but it can be clearly seen that with infinite patience, infinite skill and infinite time it could be continued for ever.

Apart from their fractal qualities a feature of fractal cuts is that they are made from a single piece of paper with nothing added and nothing taken away.

The Motif
A line divided in the proportion 2:1 with the larger portion converted into a pop-up square pyramid.

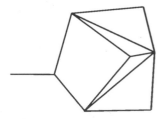

The Generating Rule
Replace each valley fold of the pyramid with a complete motif one third of the size.

Cover for Pop-up Card 9

Remove this page from the book and then cut out the cover precisely. Score along the central line and then glue the flaps of the fractal cut from page 67 into the marked positions.

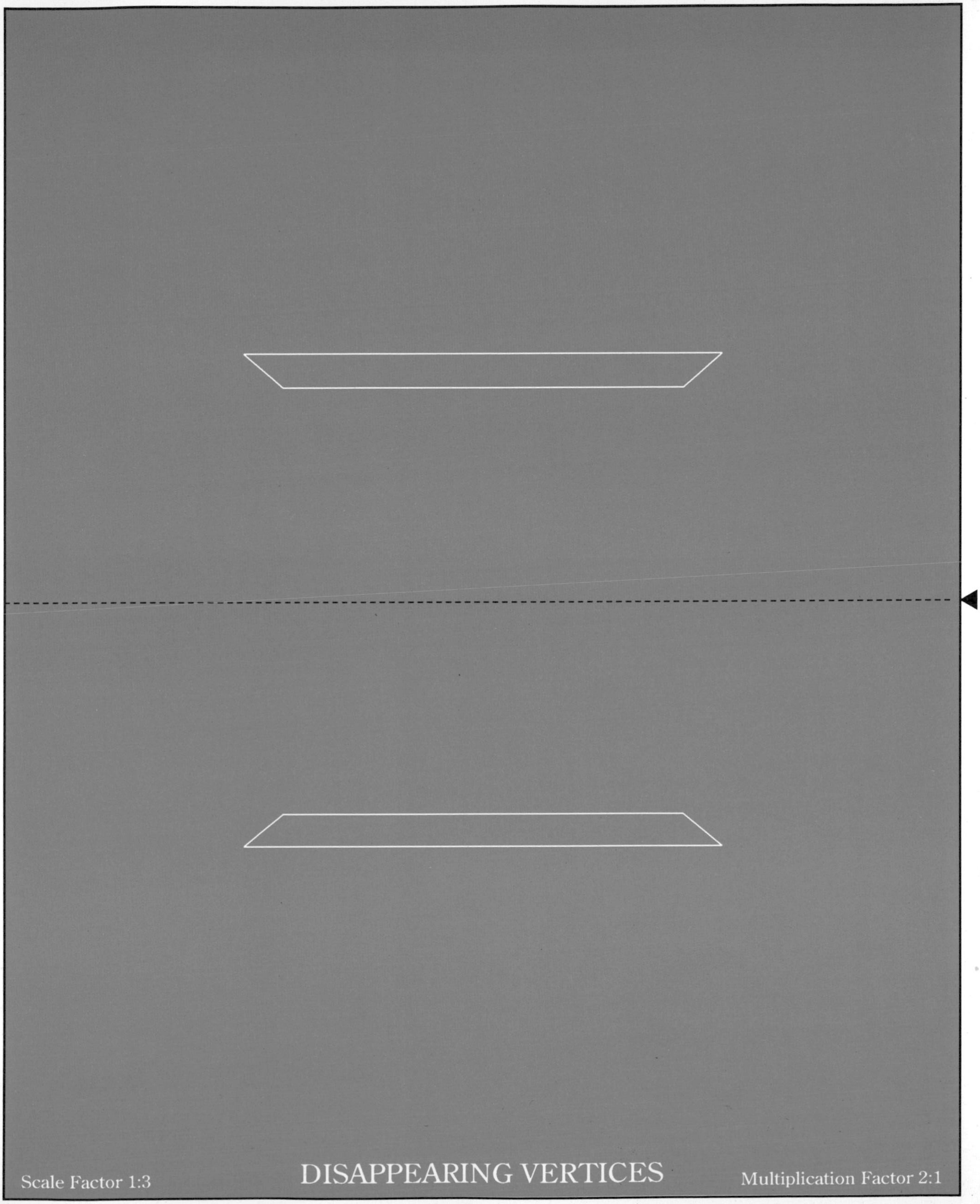

Scale Factor 1:3

DISAPPEARING VERTICES

Multiplication Factor 2:1

FRACTAL CUTS

Disappearing Vertices

A fractal cut is an endlessly developing pop-up design which belongs to the class of curious geometrical objects called fractals. Starting with a motif and applying a generating rule generation by generation, the design develops more and more detail.

This card only shows two generations, but it can be clearly seen that with infinite patience, infinite skill and infinite time it could be continued for ever.

Apart from their fractal qualities a feature of fractal cuts is that they are made from a single piece of paper with nothing added and nothing taken away.

The Motif
A sequence of folds and cuts to modify the vertex of a pyramid

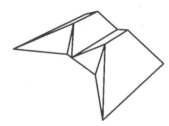

The Generating Rule
Replace each pyramid vertex with a complete motif one third of the size.

Cover for Pop-up Card 10

Remove this page from the book and then cut out the cover precisely. Score along the central line and then glue the grey area of the fractal cut from page 73 into the marked position, working outwards from the central fold.

Scale Factor 1:2   HANOI ARÊTE   Multiplication Factor 2:1

Hanoi Arête

A fractal cut is an endlessly developing pop-up design which belongs to the class of curious geometrical objects called fractals. Starting with a motif and applying a generating rule generation by generation, the design develops more and more detail.

This card only shows four generations, but it can be clearly seen that with infinite patience, infinite skill and infinite time it could be continued for ever.

Apart from their fractal qualities a feature of fractal cuts is that they are made from a single piece of paper with nothing added and nothing taken away.

The Motif
A line divided into quarters with the central two quarters converted into a pop-up square pyramid.

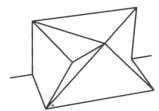

The Generating Rule
Replace each valley fold of the pyramid which is parallel to the original fold by a complete motif one half of the size.